Springer Tracts in Modern Physics 85

W0043975

Springer Tracts in Modern Physics

Volume 66   **Quantum Statistics in Optics and Solid-State Physics**
With contributions by R. Graham, F. Haake

Volume 67   **Conformal Algebra in Space-Time and Operator Product Expansion**
By S. Ferrara, R. Gatto, A. F. Grillo

Volume 68   **Solid-State Physics**   With contributions by D. Bäuerle,
J. Behringer, D. Schmid

Volume 69   **Astrophysics**   With contributions by G. Börner, J. Stewart,
M. Walker

Volume 70   **Quantum Statistical Theories of Spontaneous Emission and their
Relation to Other Approaches**   By G. S. Agarwal

Volume 71   **Nuclear Physics**   With contributions by J. S. Levinger, P. Singer,
H. Überall

Volume 72   **Van der Waals Attraction:** Theory of Van der Waals Attraction
By D. Langbein

Volume 73   **Excitons at High Density**   Edited by H. Haken, S. Nikitine. With
contributions by V. S. Bagaev, J. Biellmann, A. Bivas, J. Goll,
M. Grosmann, J. B. Grun, H. Haken, E. Hanamura, R. Levy,
H. Mahr, S. Niktine, B. V. Novikov, E. I. Rashba, T. M. Rice,
A. A. Rogachev, A. Schenzle, K. L. Shaklee

Volume 74   **Solid-State Physics**   With contributions by G. Bauer, G. Borstel,
H. J. Falge, A. Otto

Volume 75   **Light Scattering by Phonon-Polaritons**   By R. Claus, L. Merten,
J. Brandmüller

Volume 76   **Irreversible Properties of Type II Superconductors**
By H. Ullmaier

Volume 77   **Surface Physics**   With contributions by K. Müller, P. Wißmann

Volume 78   **Solid-State Physics**   With contributions by R. Dornhaus,
G. Nimtz, W. Richter

Volume 79   **Elementary Particle Physics**   With contributions by E. Paul,
H. Rollnick, P. Stichel

Volume 80   **Neutron Physics**   With contributions by L. Koester, A. Steyerl

Volume 81   **Point Defects in Metals I:** Introduction to the Theory
By N. Breuer, G. Leibfried

Volume 82   **Electronic Structure of Noble Metals, and Polariton-Mediated
Light Scattering**   With contributions by B. Bendow, B. Lengeler

Volume 83   **Electroproduction at Low Energy and Hadron Form Factors**
By E. Amaldi, S. P. Fubini, G. Furlan

Volume 84   **Collective Ion Acceleration**   With contributions by C. L. Olson,
U. Schumacher

Volume 85   **Solid Surface Physics**   With contributions by J. Hölzl,
F. K. Schulte, H. Wagner

Solid Surface Physics

Contributions by
J. Hölzl F. K. Schulte H. Wagner

With 102 Figures

Springer-Verlag Berlin Heidelberg GmbH 1979

Professor Dr. Josef Hölzl

University of Kassel (Gesamthochschule), Fachbereich Physik
Heinrich-Plett-Strasse 40, D-3500 Kassel, Fed. Rep. of Germany

Dr. Franz K. Schulte

Sektion Physik der Universität München, Lehrstuhl Professor H. Bross
Theresienstrasse 37, D-8000 München 2, Fed. Rep. of Germany
New address: Physik-Department T 30 der Technischen Universität München
James Franck Strasse, D-8046 Garching, Fed. Rep. of Germany

Dr. Heribert Wagner

Institut für Grenzflächenforschung und Vakuumphysik der Kernforschungsanlage
Jülich GmbH, Postfach 1913, D-5170 Jülich 1, Fed. Rep. of Germany

---

Manuscripts for publication should be addressed to:

Gerhard Höhler

Institut für Theoretische Kernphysik der Universität Karlsruhe
Postfach 6380, D-7500 Karlsruhe 1, Fed. Rep. of Germany

*Proofs and all correspondence concerning papers in the process of publication
should be addressed to:*

Ernst A. Niekisch

Institut für Grenzflächenforschung und Vakuumphysik der Kernforschungsanlage
Jülich GmbH, Postfach 1913, D-5170 Jülich 1, Fed. Rep. of Germany

---

ISBN 978-3-662-15809-8   ISBN 978-3-540-35253-2 (eBook)
DOI 10.1007/978-3-540-35253-2

Contents

Work Function of Metals

By *J. Hölzl* and *F.K. Schulte*. With 67 Figures

1. Introduction 1

2. Theory 2

   2.1 Definition of the Work Function 4

   2.2 Work Function of Pure Metals with Clean Surfaces 6
      2.2.1 Qualitative Discussion of Potentials and Energies
            Near a Metal Surface 6
      2.2.2 Density-Functional Formalism and Work Function 10
      2.2.3 Bulk and Surface Part of the Work Function 14
      2.2.4 Uniform-Background Model and Its Extensions 18
            a) Lang-Kohn Theory 18
            b) Extensions of the Lang-Kohn Theory 21
            c) Thin Metal Films 23
            d) Qualitative Discussion of the Anisotropy of the
               Work Function 24
      2.2.5 Wave-Mechanical Calculations for Lattice Potentials 26

   2.3 Work Function Changes Induced by Adsorbates on Pure Metals 29
      2.3.1 Classical Model 29
      2.3.2 Quantum-Mechanical Model 30
      2.3.3 Newns-Anderson Formalism 33
      2.3.4 Applications of the Newns-Anderson Formalism 34
            a) Adatom Energy Level Shift 34
            b) Results and Discussion 35
      2.3.5 Tight-Binding Approximation 37
      2.3.6 Applications of the Density-Functional Formalism 38
            a) A Single Adatom 38
            b) Representation of the Adsorbate Layer by a Charge Slab 39

   2.4 Work Function of Alloys 42

3. Experimental Procedures 45

  3.1 Survey of Experimental Methods 45
  3.2 Absolute Methods 45
      3.2.1 Thermionic Emission 45
      3.2.2 Photoelectric Method 49
      3.2.3 Field Emission 50
  3.3 Relative Methods 51
      3.3.1 Diode Methods and Examples for Practical Configuration 52
      3.3.2 Condenser Methods 58
          a) Vibrating Capacitor Methods 59
          b) Systematic Sources of Error 61
          c) Detection System 62
          d) Form of the Kelvin Method in Practice 63
          e) Other Condenser Methods 67

4. Work Function of Pure Metals with Clean Surfaces 68

  4.1 Summary of Theoretical Models Used for the Calculation of
      the Work Function of Pure Metals with Clean Surfaces 69
      4.1.1 Empirical and Semiempirical Studies 69
      4.1.2 Outline of Quantum-Mechanical Treatments 72
  4.2 Preparational Procedures 73
  4.3 Temperature Effects on the Work Function 77
  4.4 Mechanical Stress Dependence of Work Function 80
  4.5 Compilation of Work Function Data on Pure Metals 85

5. Work Function Changes Induced by Adsorbates on Pure Metals 85

  5.1 Work Function as a Measure of Coverage 96
      5.1.1 Calibration Methods 96
      5.1.2 Use of the Work Function as a Measure of Coverage 98
          a) Thermodynamics 99
          b) Surface Kinetics 100
          c) Surface Diffusion 102
  5.2 Work Function and Representative Surface Data of Adsorbate Systems .. 104
      5.2.1 Static Substrate Model 104
          a) Theoretical Relationships and Basic Adsorption Experiments 104
          b) Experiments Relating to Gurney's "Depolarization Model" ... 109
          c) Dependence of $\Delta\Phi(\theta)$ on the Structure of the
             Substrate Surface 112
      5.2.2 Dynamic Substrate Model 113
  5.3 Compilation of Ad-Systems Connected with Work Function Studies 116

6. Work Function of Alloys 126

  6.1 Summary of Theoretical Treatments of Alloy Systems 126
  6.2 Preparational Procedure and Usefulness of Concentration Graphs 127
  6.3 Work Function and Surface Composition of Alloys 130
  6.4 Work Function and Other Alloy Characteristics 133
      6.4.1 Work Function and Bulk/Surface Properties 133
      6.4.2 Use of Work Function Measurement for Obtaining Thin Alloy
            Film Diffusion Parameters 136

References 140

Physical and Chemical Properties of Stepped Surfaces

By *H. Wagner*. With 35 Figures

1. Introduction 151

2. Characterization of Stepped Surfaces 153
  2.1 Surface Crystallography 153
  2.2 Experimental Evidence for Step Structures 156
      2.2.1 Field Ion Microscopy 157
      2.2.2 Low Energy Electron Diffraction (LEED) 159
      2.2.3 LEED from Stepped Surfaces 161
      2.2.4 Electron Microscopy 168

3. Properties of Clean Stepped Surfaces 169
  3.1 Structural Properties 169
  3.2 Thermal Stability of Step Structures 173
      3.2.1 Experimental Observations 173
      3.2.2 Theoretical Considerations 175
  3.3 Electronic Properties 182
      3.3.1 Work Function 182
      3.3.2 Surface States 188

4. Interaction of Atoms and Molecules with Stepped Surfaces 190
  4.1 Adsorption Kinetics 191
  4.2 Adsorption States 196
  4.3 Adsorbate Structures 200
  4.4 Catalytic Reactions 202
  4.5 Atom and Molecule Scattering from Stepped Surfaces 209
  4.6 Surface Diffusion 213

5. Conclusions 217

References 219

Work Function of Metals

J. Hölzl and F. K. Schulte

1. Introduction

F.K. Schulte and J. Hölzl

The work function (WF) of a metal can be defined as the minimum energy required
to extract one electron from a metal. Obviously the WF is one of the fundamental
electronic properties of bare and coated metallic surfaces.

In the first half of this century the WF was discussed mainly in connection
with the thermionic emission of electrons, which was first observed by EDISON in
1884. RICHARDSON (1901) and DUSHMAN (1923) derived the equation for the thermionic
emission current. In the 1930s WIGNER and BARDEEN did a calculation to determine
the WF of simple metals which is still the basis for modern theoretical treatments.
In an excellent article HERRING and NICHOLS /1.1/ reviewed the field of thermionic
emission up to 1949.

Since 1949 the interest in the WF has been stimulated by ultra-high-vacuum
technique, by new experimental surface techniques and by technological interest
in thermionic converters and in catalysis. As a result, progress both theoretical
and experimental, is now very rapid, and a series of review papers has been published.

FOMENKO (1966, 1970) /1.2,3/ tabulated measured WF data for many elements and
compounds and recommended selected values. A more detailed critical review of WF
measurements and results for elements, alloys, and compounds is due to RIVIERE
(1969) /1.4/. The most recent compilation of selected WF data for elements is contained in a paper by MICHAELSON (1977) /1.5/.

With a view on thermionic emission HAAS and THOMAS (1972) /1.6/ reviewed measurement methods and the theory of the WF up to 1972. Applications of the density-
-functional formalism to calculations of the WF have been reviewed by LANG /1.7/.

In the subsequent article our main interest is focused on the most recent theoretical and experimental studies on the WF of metals.

The theory (Chap. 2) concentrates on calculations of the WF of metals. It contains studies on surfaces of pure metals without and with adsorbates and a short
section on alloys. It does not contain the thermodynamics of electron emission or
a discussion of non uniform, "patchy" surfaces of polycrystalline metals. For these
topics comprehensive reviews are already available /1.1,6/.

In Chap. 3 the experimental procedures are described. Therein absolute and relative methods are reviewed.

Since there is a comprehensive presentation dealing with absolute methods /1.4/,
relative methods are discussed in more detail.

In the following chapters experimental results for pure metals with clean surfaces (Chap. 4), for pure metals with adsorbates (Chap. 5), and for alloys (Chap. 6)
are reviewed. For the reader mainly interested in experimental results some short
theoretical guidelines are also incorporated in these chapters. The main interest
of the author (J.H.) is a demonstration of how far the WF experiments can contribute to a variety of surface studies rather than giving a critical discussion of
experimental techniques.

In Tables 4.3 and 5.3 the reader can find a compilation of the WF data for clean
surfaces and surfaces with adsorbates respectively.[1] These data have been brought
together without any attempt at critical selection. As far as possible they represent the most recent measurements. Where different values are tabulated for the
same surface the reader is invited to consult the original papers for purpose of
intercomparison.

2. Theory

F.K. Schulte

For the definition of the WF thermodynamic concepts are used and no reference is
made to a single-particle model. Throughout this chapter the term WF always denotes the so-called "true WF" which is defined for uniform (not patchy) surfaces
without reference to any particular experiment. As will be discussed in Chap. 3
precautions are sometimes necessary when making comparisons with measured WF's.

WF's of pure metals with clean surfaces are dealt with in Sect. 2.2. The reader
interested in empirical and semiempirical studies that correlate the WF's of pure
metals with a series of bulk and surface properties is referred to Sect. 4.1. In
that section the correlation with electronegativities established by GORDY and
THOMAS /2.1/ and by STEINER and GYFTOPOULOS /2.2/ is discussed in some detail.
Empirical studies do not, of course, provide deep insight into the physical nature
of the WF. They can, however, be used to predict the WF's of many more materials
and crystal faces than have been studied in first principles calculations.

---

[1] The compilation of the various results listed in these tables was much facilitated by the kind cooperation of the "Zentralstelle für Dokumentation" (Karlsruhe, Germany). The authors are greatly indebted to the staff of that office.

To reveal the physical factors that determine the WF a very simple single-particle picture is sketched (Sect. 2.2.1). Its basic features are due to the pioneering work by WIGNER and BARDEEN /2.3,4/. This picture forms the basis for practically all modern WF calculations.

It is formally justified in terms of the density-functional formalism developed by HOHENBERG, KOHN and SHAM /2.5-7/ (Sect. 2.2.2). Since most of the modern WF calculations employ the density-functional formalism, a short presentation of its main features is included.

Calculations of the bulk contribution of the WF's of nontransition and transition metals (recent calculations are due to HODGES and collaborators /2.8-10/) are summarized in Sect. 2.2.3. In this connection several alternative definitions of the bulk contribution are discussed. The uncertainties in the calculated bulk contributions of the WF's are still relatively large, ranging in some cases over 1 eV.

Many of the calculations of total WF's are based on the uniform-background or jellium model, which together with its extensions is discussed in Sect. 2.2.4. The most complete and most exact investigation employing the uniform-background model ist due to LANG and KOHN /2.11,12/. For s-p-bonded metals it has led to excellent agreement with experiments. Unfortunately it was not possible to extend these calculations to transition metals. A uniform-background model with a corrugated surface has been studied by SMOLUCHOWSKI /2.13/. His prediction that higher WF's correlate with closer packed (smoother) surfaces is in general agreement with experiments.

A few more recent calculations which start from a lattice potential, are discussed in Sect. 2.2.5. The results of these most sophisticated calculations deviate in some cases rather markedly (by up to 1 eV) from measured WF's.

WF changes induced by adsorbates on pure metals are dealt with in Sect. 2.3. The theory of these WF changes can be traced back to the classical model due to LANGMUIR /2.14/ (Sect. 2.3.1) and to the quantum-mechanical treatment of GURNEY /2.15/ (Sect. 2.3.2). Based on their concepts, quantitative studies have been performed with the help of the Newns-Anderson formalism /2.16/. In its simplest form this formalism is described in Sect. 2.3.3. Calculations have been done almost exclusively for alkali adsorbates on an idealized flat metal surface. The results derived for the initial slopes of the WF vs. coverage curves fit the experimental results very well. Moreover, by fitting the WF vs. coverage curves, valuable information has been obtained concerning the position and the width of the valence energy level (Sect. 2.3.4).

Recent applications of the tight-binding approximation to calculate the WF for disordered and ordered alkali layers adsorbed on s-p-bonded substrates /2.17,18/ are discussed in Sect. 2.3.5.

Applications of the density-functional formalism to the adsorption problem are dealt with in Sect. 2.3.6. Dipole moments calculated for single adatoms (alkalis, H, and O) on a uniform-background substrate agree satisfactorily with dipole moments derived from measured initial slopes. To include finite coverages, the ion cores of alkali adsorbates have been replaced by a uniform slab of background charge. The WF minima calculated for this slab model fit the measurements remarkably well.

Phenomenological studies of WF changes /2.19-22/, which employ a variety of different concepts, are left out of consideration. The reader interested in these treatments is referred to the original papers.

Sect. 2.4 deals mainly with a very interesting paper on the WF of alloys, which is due to GELATT and EHRENREICH /2.23/. This work seems to be well suited as starting point for further investigations.

2.1 Definition of the Work Function

In the introduction the WF has been defined loosely as the minimum energy required to extract one electron from a metal. In this definition the final state of the electron needs to be specified. For the energy to be a minimum the electron must be finally at rest. Then the WF still depends on the final position of the electron (see also /2.24/). If a finite crystal is considered and the final position is chosen at an infinitely large distance from the surface, it is not possible to discriminate between WF's of different crystal faces. For the definition of the WF

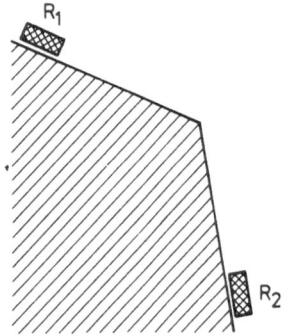

Fig. 2.1. Sketch of final positions R_1 and R_2 of removed electrons for the definition of the WF's of different monocrystal faces. See text

of a clean monocrystal face the distance of the electron from the face should be
so large that the image force is negligible (typically 10^{-4} cm) but it should be
small compared with the distance from another face with a different WF. Fig. 2.1
shows an edge of a monocrystal and regions R_1 and R_2 which fulfill these requirements. If the face is covered by an adsorbate the distance of the electron from
the surface under consideration must also be large compared with the distance between the adatoms.

A difference between the WF's of the adjacent faces in Fig. 2.1 implies a potential difference between R_1 and R_2. For this reason, outside a monocrystal whose
surfaces have different WF's, there exists a macroscopic field.

For zero temperature the definition of the WF can be made more precise in the
following way (see also /2.25/); the WF is the energy difference between two states
of the whole crystal. In the initial state the neutral crystal containing N electrons
is assumed to be in its ground state with energy E_N. In the final state one electron
is removed from the crystal to a region specified in Fig. 2.1. There it is assumed
to be at rest and accordingly has only electrostatic energy denoted by ϕ_V. The crystal with the remaining N-1 electrons is assumed to be in its ground state with
energy E_{N-1}. Combining all the energies we obtain

$$\phi = (E_{N-1} + \phi_V) - E_N. \tag{2.1}$$

For temperatures greater than zero the removal of an electron from the metal is
to be considered as a thermodynamic change of state. The difference $E_N - E_{N-1}$ has to
be replaced by the derivative of the Helmholtz free energy F with respect to the
electron number N, whereby the temperature T and the volume V are kept constant.
This derivative is the electrochemical potential[1] of the electrons, μ

$$E_N - E_{N-1} \rightarrow \left(\frac{\partial F}{\partial N}\right)_{T,V} = \mu. \tag{2.2}$$

Thus we obtain the generalization of (2.1) for nonzero temperatures

$$\phi = \phi_V - \mu. \tag{2.3}$$

---

[1] Throughout this chapter no linguistic difference is made between the terms potential and potential energy of an electron.

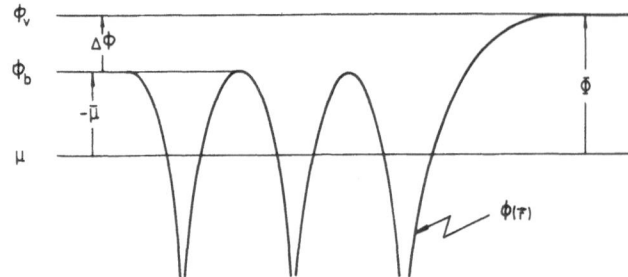

<u>Fig. 2.2.</u> Schematic plot of various energies relevant to the definition of the WF. See text

If an energy level ϕ_b, which is characteristic of the electrostatic potential in the bulk of the metal, is used as a reference level, [2] the WF can be subdivided into a surface-dependent part $\Delta\phi = \phi_v - \phi_b$ and the chemical potential $\bar{\mu} = \mu - \phi_v$, which depends on bulk properties only, so that

$$\Phi = \Delta\phi - \bar{\mu}. \tag{2.4}$$

For a schematic representation of the electrostatic potential $\phi(r)$ near a metal surface, together with the energies relevant to the definition of the WF, see Fig. 2.2.

Note that in the definition of the WF given in this section no single-particle model has been employed. A qualitative discussion of the WF in a very simple single-particle picture will be given in Sect. 2.2.1.

2.2 Work Function of Pure Metals with Clean Surfaces

2.2.1 Qualitative Discussion of Potentials and Energies Near a Metal Surface

In this subsection the basic factors determining the WF are discussed in a qualitative way. A simple picture is presented which will be justified by the formal theory outlined in the following subsection.

---

[2] In common use are the total average of the bulk electrostatic potential and the potential average extended over the interstitial regions between the atoms only. For details see Sect. 2.2.3.

An important quantity which determines the WF is the potential of an electron near the surface. According to electrostatic theory an electron feels the electro-static or Hartree potential

$$\phi(\underline{r}) = -e \int \frac{\rho_{tot}(\underline{r}')}{|\underline{r}-\underline{r}'|} d\underline{r}', \qquad (2.5)$$

where ρ_{tot} is the total charge density of the crystal. The dependence of ϕ on a coordinate perpendicular to the surface is schematically shown by the dashed curve in Fig. 2.3.

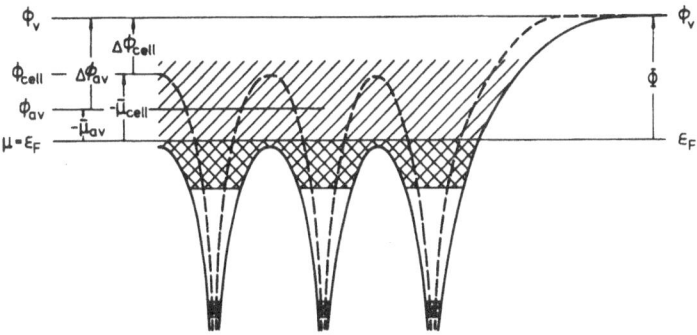

Fig. 2.3. Potentials and energies near a metal surface. See text

In (2.5) we have not taken into account any correlations between the individual electrons. However, the electrons tend to stay away from one another for two rea-sons /2.26/: the Pauli principle does not allow two electrons with the same spin to be at the same place, and the electrons repel each other via the Coulomb inter-action. These mechanisms give rise to the exchange and correlation potential, re-spectively. A rough estimate of the exchange and correlation potential is provided by considering an electron at three different positions:

(i) The electron is assumed to be well outside the metal. Compare Fig. 2.4a. In that situation it repels electrons at the metal surface. Thus the charge density assumed in (2.5) is modified. The difference consists of a surface charge density indicated by the hatching in Fig. 2.4a. This positive surface charge density at-metal   vacuum

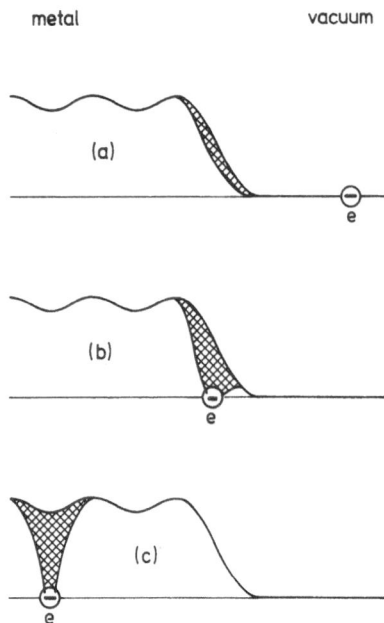

Fig. 2.4. Charge distribution due to exchange and correlation effects. A single
electron e is assumed to be (a) well outside the metal, (b) in the surface region,
(c) in the metal interior. See text

tracts the electron. The corresponding potential is well known from elementary
electrostatics, and is called the image potential, $-e^2/4x$, where x denotes the
distance from the surface charge.

(ii) The concept of an image charge clearly breaks down when the electron
merges into the surface electron density. This case is shown in Fig. 2.4b. Repulsion of neighboring electrons results in a positive hole surrounding an electron
in the surface region. Since the available electron charge density is larger on
the metallic side of this electron, the center of gravity of the charge hole is
shifted from the electron position toward the metal interior. Therefore the positive
hole attracts the electron towards the metal interior. Consequently the potential
decreases in the -x direction.

(iii) If the electron is well inside the metal (Fig. 2.4c) the positive hole
surrounding it becomes approximately spherically symmetric. For simplicity let us

assume that the charge density ρ_h of the hole around an electron at $\underline{r}$ is given by

$$\rho_h(\underline{r}') = \begin{cases} en(\underline{r}), & |\underline{r}'-\underline{r}| \leq r'_s \\ 0, & |\underline{r}'-\underline{r}| > r'_s \end{cases}, \tag{2.6}$$

where n denotes the unperturbed electron density. Since the total hole charge must be e, r'_s is the so-called Wigner-Seitz radius, which is related to the electron density $n(\underline{r})$ via the equation $(4\pi/3)r'^3_s = n(\underline{r})^{-1}$. Now it is easily shown that the potential of the electron in the field of the hole charge (2.6) is $-cn(\underline{r})^{1/3}$ with a positive constant c. For typical metallic densities it is of the order of some eV, which clearly demonstrates the importance of the exchange and correlation potential to the WF.

Combining these results, the exchange and correlation potential, which we denote as v_{xc}, is identical to the image potential outside the metal. It decreases continuously in the surface region, and varies with the electron density as $-cn(\underline{r})^{1/3}$ inside the metal. If we add the exchange and correlation potential to the electrostatic potential we obtain the effective potential

$$v_{eff}(\underline{r}) = \phi(\underline{r}) + v_{xc}(\underline{r}), \tag{2.7}$$

which is also shown in Fig. 2.3.

The electrons move in this effective potential $v_{eff}(\underline{r})$. The eigenstates are solutions of a Schrödinger equation with the potential $v_{eff}(\underline{r})$. Due to the periodicity of the potential the allowed energies of extended bulk states lie in energy bands. In addition there may be bands of surface states. The amplitude of surface states is large only near the surface, and decays toward the interior of the metal. At zero temperature the electron states are occupied according to the Pauli principle up to maximum energy, the Fermi energy ε_F. In Fig. 2.3 the allowed energies are indicated by the hatched areas. Cross-hatched areas indicate occupied states. In this simple picture the WF, defined as the minimum energy required to remove an electron from the crystal, is the difference between the electrostatic potential in the vacuum region ϕ_v and the Fermi energy ε_F

$$\phi = \phi_v - \varepsilon_F. \tag{2.8}$$

A comparison with (2.3) shows that ε_F corresponds to the electrochemical potential $\mu = E_N-E_{N-1}$ at zero temperature.

Needless to say, the considerations in this subsection do not provide a method for calculating the WF. For a calculation of the WF the metal has to be considered as a many-body system. This is done in the following section. We will see there that the many-body problem can be reduced to a one-body form, which corresponds closely to the picture just outlined.

2.2.2 Density-Functional Formalism and Work Function

Most of the modern computations of the WF are based on the density-functional formalism, which has been developed by HOHENBERG, KOHN and SHAM /2.5-7/. We want to present only the main features of this formalism. The reader interested in more details is referred to the excellent review article by LANG /2.12/.

In the density-functional formalism the conduction electrons (electrons) are viewed as an interacting electron gas in the external potential $v(\underline{r})$ caused by the ion cores (atomic nuclei). As its name suggests the central quantity in the density--functional formalism is a density functional. This density functional, $E_v[n]$, has been defined by HOHENBERG and KOHN /2.5/ as

$$E_v[n] = e \int v(\underline{r})n(\underline{r})d\underline{r} + F[n],$$ (2.9)

where the functional $F[n] = \langle \Psi_n|T+U|\Psi_n\rangle$ is the expectation value of the total kinetic and interaction energy in the ground state of an electron system with density $n(\underline{r})$. $n(\underline{r})$ is not necessarily the ground-state density of electrons in the external poten- tial $v(\underline{r})$. HOHENBERG and KOHN showed, however, that the functional $E_v[n]$ is minimum for the ground-state density $n(\underline{r})$, if the subsidiary condition $\int n(\underline{r})d\underline{r} = N$ is taken into consideration. The minimum of $E_v[n]$ is the ground-state energy.

To determine the ground-state properties of the electron system, KOHN and SHAM /2.6/ assumed that there is a fictitious system of noninteracting Fermions with the same ground-state density as in the interacting electron system. This density is then given by

$$n(\underline{r}) = \sum_{i=1}^{N} |\psi_i(\underline{r})|^2,$$ (2.10)

where the summation is to be extended over the lowest lying one-particle states ψ_i of the noninteracting system. The kinetic energy of the electrons can be approx- imated by

$$T_s[n] = -\frac{\hbar^2}{2m} \sum_{i=1}^{N} \int \psi_i^*(\underline{r})\Delta\psi_i(\underline{r})d\underline{r}.$$ (2.11)

If in addition the Hartree part of the electron interaction energy is written explicitly and the rest is denoted as exchange and correlation energy, E_{xc}, we obtain

$$F[n] = T_s[n] + \frac{e^2}{2} \int \frac{n(\underline{r})n(\underline{r}')}{|\underline{r}-\underline{r}'|} \, d\underline{r}d\underline{r}' + E_{xc}[n]. \tag{2.12}$$

For applications of the density-functional formalism it is convenient to perform the following gradient expansion

$$G[n] = T_s[n] + E_{xc}[n]$$

$$= \int d\underline{r} \, [g^{(0)}(n(\underline{r})] + g^{(2)}(n(\underline{r}))|\nabla n(\underline{r})|^2 + \ldots]. \tag{2.13}$$

$g^{(0)}$ is the energy per unit volume in a homogeneous electron gas with density n. It is given by

$$g^{(0)}(n) = \varepsilon^{(0)}(n)n, \text{ with } \varepsilon^{(0)}(n) = t^{(0)}(n) + \varepsilon_{xc}^{(0)}(n), \tag{2.14}$$

where $t^{(0)}$ and $\varepsilon_{xc}^{(0)}$ denote the mean kinetic and exchange and correlation energy, respectively, per particle, i.e.,

$$t^{(0)}(n) = 0.3 \, (\hbar^2/m)(3\pi^2 n)^{2/3}, \tag{2.15}$$

$$\varepsilon_{xc}^{(0)}(n) = -0.75e^2(3n/\pi)^{1/3} + \varepsilon_c^{(0)}(n). \tag{2.16}$$

The correlation energy, $\varepsilon_c^{(0)}$, has been extensively studied in the literature /2.26-30/. In applications of the density-functional formalism WIGNER's formula /2.29/

$$\varepsilon_c^{(0)}(n) = -0.44e^2/(r_s+7.8) \tag{2.17}$$

is often used. r_s denotes the dimensionless Wigner-Seitz radius in units of Bohr's hydrogen radius a_0: $r_s = (3/4\pi n)^{1/3}/a_0$. Another more recent expression for the correlation energy is due to GUNNARSSON, et al. /2.30/.

Just as for $g^{(0)}$, $g^{(2)}$ also contains contributions from the kinetic energy T_s and from the exchange and correlation energy E_{xc}

$$g^{(2)}(n) = t^{(2)}(n) + \varepsilon_{xc}^{(2)}(n). \tag{2.18}$$

The contribution to the kinetic energy $t^{(2)}$ can be found from an RPA dielectric constant /2.31/

$$t^{(2)}(n) = \frac{\hbar^2}{72mn}. \tag{2.19}$$

The coefficient $\varepsilon_{xc}^{(2)}$ has only recently been numerically determined by RASOLT and GELDART /2.32/.

Two schemes for practical calculation of ground-state properties result when the minimum condition for the functional

$$E_v[n] = e\int v(\underline{r})n(\underline{r})d\underline{r} + \frac{e^2}{2}\int\frac{n(\underline{r})n(\underline{r}')}{|\underline{r}-\underline{r}'|}\,d\underline{r}d\underline{r}' + G[n] \qquad (2.20)$$

is explicitly written.

If the gradient expansion (2.13) for $G[n]$ is employed, there results

$$\phi(\underline{r}) + g^{(0)'}(n(\underline{r})) - g^{(2)'}(n(\underline{r}))|\nabla n(\underline{r})|^2 - 2g^{(2)}(n(\underline{r}))\nabla^2 n(\underline{r}) + \ldots = \mu, \qquad (2.21)$$

where μ is constant and ϕ is the electrostatic potential. In terms of the external potential v it is given by

$$\phi(r) = v(r) + e^2\int\frac{n(\underline{r}')}{|\underline{r}-\underline{r}'|}\,d\underline{r}'. \qquad (2.22)$$

By the approximations $g^{(0)} = t^{(0)}$ and $g^{(2)} = 0$, (2.21) reduces to the Thomas-Fermi equation.

If the kinetic energy T_s is given exactly by (2.11), and a gradient expansion analogous to (2.13) is performed for the exchange and correlation energy, the states ψ_i can be varied to minimize the functional E_v. In this way there result the Schrödinger equations

$$\left[-\frac{\hbar^2}{2m}\Delta + v_{eff}(\underline{r})\right]\psi_i(\underline{r}) = \varepsilon_i\psi_i(\underline{r}), \qquad (2.23)$$

with an effective potential

$$v_{eff}(\underline{r}) = \phi(\underline{r}) + v_{xc}(\underline{r}). \qquad (2.24)$$

$\phi(\underline{r})$ is the electrostatic potential, [cf. (2.5,22)], while the exchange and correlation potential is given by

$$v_{xc}(\underline{r}) = \frac{\delta E_{xc}[n]}{\delta n(\underline{r})} = v_{xc}^{(0)}(\underline{r}) + v_{xc}^{(2)}(\underline{r}) + \ldots, \qquad (2.25a)$$

with a local term $v_{xc}^{(0)}$, which is equal to the exchange and correlation part μ_{xc} of the chemical potential of a homogeneous electron gas

$$v_{xc}^{(0)}(\underline{r}) = \mu_{xc}(n(\underline{r})) = \frac{d}{dn}(\varepsilon_{xc}^{(0)}(n)n)|_{n=n(\underline{r})}, \qquad (2.25b)$$

and with a gradient term

$$v_{xc}^{(2)}(\underline{r}) = -\varepsilon_{xc}^{(2)}{}'(n(\underline{r})))|\nabla n(\underline{r})|^2 - 2\varepsilon_{xc}^{(2)}(n(\underline{r}))\nabla^2 n(\underline{r}). \tag{2.25c}$$

Aside from v_{xc} the $x\alpha$ potential, $v_{x\alpha}$, proposed by SLATER /2.33/ has been frequently used. It is given by

$$v_{x\alpha}(\underline{r}) = 1.5\alpha v_x^{(0)}(\underline{r}), \tag{2.26}$$

where $v_x^{(0)}$ is the exchange part of $v_{xc}^{(0)}$. The $x\alpha$ potential is consistent with the exchange and correlation energy

$$E_{x\alpha}[n] = 1.5\alpha E_x[n]. \tag{2.27}$$

Only if this expression is used to determine the total energy, can the virial theorem be satisfied exactly /2.34/. $V_{x\alpha}$ is obtained when in (2.25a) E_{xc} is replaced by $E_{x\alpha}$.

Several criteria have been proposed to determine α /2.33/. The resulting values lie between 2/3 and 1. Though by all the commonly used criteria correlation effects are not systematically taken into consideration, by chosing α larger than 2/3 $v_x^{(0)}$ is corrected in the same direction as by the addition of $v_c^{(0)}$.[3] Since, however, α is generally not systematically determined to include correlation, it seems to be more appropriate to use $v_{xc}^{(0)}$ instead of $v_{x\alpha}$.

Equations (2.23-25) correspond closely to the physical picture outlined in the previous section. The density-functional formalism provides a systematic method for a determination of the exchange and correlation potential v_{xc} in (2.7).

It must be realized, however, that the energies ε_i characterize noninteracting Fermion states, and not electron or quasi-particle states. Therefore it is by no means obvious that (2.8) holds, i.e., that the difference between the vacuum potential ϕ_v and the Fermi energy ε_F of the noninteracting system can be identified with the WF. It has, however, been shown that at zero temperature ε_F is equal to the electrochemical potential μ defined in (2.2). LANG and KOHN /2.11/ proved this identity for the uniform-background or jellium model. SCHULTE /2.35/ gave a proof for the more general case of a periodic lattice potential. As a consequence the definitions (2.1) and (2.8) are equivalent if ideal surfaces are considered in the framework of the density-functional formalism.

---

[3] If we set $v_{x\alpha} = v_{xc}^{(0)}$ and make use of Wigner's formula for the correlation energy, the resulting values of α vary between 0.77 for $r_s = 2$ and 0.91 for $r_s = 6$.

Therefore for a determination of the WF it is sufficient to solve (2.23); it is not necessary to start from the more complicated Schrödinger-like equation, which results from the Dyson equation for the one-particle Green's function /2.7,27/. With practical computation of the WF in mind it must be appreciated that the effective potential has to be determined self-consistently, in other words (2.8,22-25) must be solved simultaneously. For this reason, there are, up to now, only relatively few calculations of the WF.

2.2.3 Bulk and Surface Part of the Work Function

In (2.4) the WF was subdivided into a surface dependent contribution $\Delta\phi = \phi_v - \phi_b$ and a bulk contribution, the chemical potential $\bar{\mu} = \mu - \phi_b$. In this section we will be concerned mainly with calculations of the chemical potential $\bar{\mu}$, which has often been investigated[4] /2.4,8-10,44-49/.

As discussed in the previous section, μ can be identified with the Fermi energy ε_F. Therefore the chemical potential μ can, in principle, be obtained from a bulk band structure calculation. For calculations of $\Delta\phi$, that will be discussed in the next sections, the surface electronic structure must be taken into account explicitly.

Two different reference levels ϕ_b are in common use in the literature,[5] and care must be taken not to confuse them /2.9,46/. In investigations starting from a uniform-background or jellium model the average value of the total electrostatic potential in the bulk, $\phi_b = \phi_{av}$, is used /2.12,36/. If the atomic structure of the metal is taken into account in a cellular approximation, the electrostatic potential averaged over only the interstitial regions between the atoms, $\phi_b = \phi_{cell}$, is usually adopted as reference level /2.4,8-10,44,45/. The corresponding energies will be denoted by the subscripts av and cell, respectively. For illustration $\Delta\phi_{av,cell}$ and $\bar{\mu}_{av,cell}$ are shown on the left of Fig. 2.3.

---

[4] Chemical potentials $\bar{\mu}$ are important not only as large contributions to the WF. They also determine the relative positions of the levels ϕ_b of two metals in electrical contact with each other, see, e.g., /2.36/. Also, HODGES and STOTT /2.37/ proposed that $\bar{\mu}$ should be an important parameter governing charge transfer in alloys.

Addition of $\bar{\mu}$ to the experimental value of the WF results in the dipole barrier $\Delta\phi$. Recently the dipole barrier has attracted much attention /2.37-43/, mainly in connection with the positron WF. Since for positrons the negative of $\Delta\phi$ contributes to their WF, it has been speculated that the positron WF could become negative /2.38-43/.

[5] In addition a superposition of atomic charge densities has been used as reference /2.48,49/. The surface dipole representing the relaxation of this charge distribution is small compared with $\Delta\phi_{av}$ or $\Delta\phi_{cell}$ /2.48,49/.

For the uniform-background model the chemical potential $\bar{\mu}_{av}$ can be derived from (2.23-25) together with (2.16) and WIGNER's formula (2.17) for the correlation energy.[6] It is given by

$$\bar{\mu}_{av} = \frac{\hbar^2 k_F^2}{2m} + \mu_{xc}(n) = \frac{50.1}{r_s^2} - \frac{16.6}{r_s} - \frac{12.0}{r_s+7.8}(1 + \frac{r_s}{3(r_s+7.8)}) \quad [eV], \qquad (2.28)$$

where $k_F = (3\pi^2 n)^{1/3}$ denotes the Fermi momentum.

The chemical potential $\bar{\mu}_{cell}$ was first investigated by WIGNER and BARDEEN /2.3/. They expressed a large part of $\bar{\mu}_{cell}$ by the cohesive energy, thus explaining the strong correlation between the cohesive energy and the WF, which was first pointed out by SOMMERFELD /2.50/. Since there are excellent reviews /2.44,45/,[7] we will not deal further with this aspect of the chemical potential, but only quote the result

$$\bar{\mu}_{cell} = \varepsilon_{coh} + \varepsilon_{ion} + \frac{2}{3}t^{(0)}(\bar{n}) + \mu_{xc}(\bar{n}) - \varepsilon_{xc}(\bar{n}) + \varepsilon_{coul} . \qquad (2.29)$$

ε_{coh} denotes the cohesive energy of the metal and ε_{ion} the ionization energy of the free metal atoms. $\bar{n}$ is the mean electron density and $\varepsilon_{coul} = 0.6Ze^2/R$ is the electrostatic self-energy per electron if the Z conduction electrons are assumed to be uniformly distributed over the Wigner-Seitz sphere of radius R.

Equations (2.28,29) can be used for a rough estimate of the chemical potentials for simple free electron metals. We shall next discuss more refined methods for the determination of $\bar{\mu}_{cell}$ that have been applied recently /2.8-10/.

For nontransition metals HEINE and HODGES /2.8/ computed corrections to the chemical potential $\bar{\mu}_{cell}$ given in (2.29). In the framework of pseudopotential theory they took into consideration the influences of the effective mass and of the non-uniformity of the conduction electron gas. The corresponding corrections range up to 1.2 eV. The results are shown in Fig. 2.5a.

NIEMINEN and HODGES /2.10/ determined $\bar{\mu}_{cell}$ from band structure calculations. From suitable band structures they took the difference between the Fermi level ε_F and the muffin-tin zero v_0. Since v_0 is, according to (2.24), composed of the electrostatic potential ϕ_{cell} and an exchange and correlation potential, a corresponding correction was made to eliminate the exchange and correlation part. In this way results were obtained for transition metals (see Fig. 2.5b) for which simpler models do not apply.

---

[6] If the correlation part of the chemical potential, μ_c, is expressed by the formula due to GUNNARSSON et al. /2.30/, the difference from Wigner's formula is remarkable. In the metallic density range ($r_s = 2$-6) it amounts up to 0.4 eV (for $r_s = 2$).

[7] For a modern form of these calculations see also /2.8/.

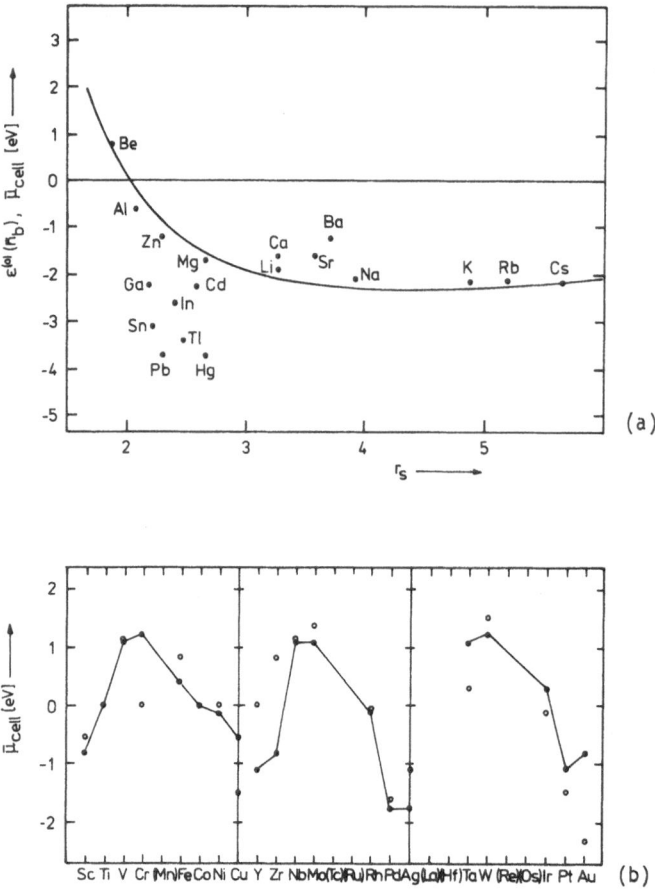

Fig. 2.5. Bulk parts of the WF's for (a) nontransition metals, (b) transition metals.
(a) Comparison between the energy per electron $\epsilon^{(0)}(\bar{n}_b)$ plotted as function of r_s =
$(4\pi\bar{n}_b/3)^{-1/3}/a_0$ and the chemical potentials $\bar{\mu}_{cell}$ calculated by HEINE and HODGES /2.8/.
(Based on /2.9/). (b) Comparison between the chemical potentials $\bar{\mu}_{cell}$ calculated from
a pressure cell-boundary relation (points) and those derived from band structures
(circles). The elements in parentheses have not been considered. (Based on /2.10/)

An alternative way of estimating $\bar{\mu}_{cell}$ has been proposed by HODGES and NIEMINEN
/2.9,10/. They make use of a pressure cell boundary relation, which results if the
energy change of the cell energy E_{cell} with the cell volume Ω, $\delta E_{cell}(\Omega) = -pd\Omega$,
is written explicitly, where the cell energy E_{cell} is defined by (2.20) and (2.13)
with the integrations extended over a single Wigner-Seitz cell only.

For nontransition metals HODGES /2.9/ adopted the gradient expansion in (2.13). He found that for gradient terms with $g^{(2)}(n) \sim n^{-1}$ - as proposed by VON WEIZSÄCKER /2.51/ and as given in (2.19) - the pressure cell boundary relation has the simple form $p = \bar{n}_b(\bar{\mu}_{cell} - \varepsilon^{(0)}(\bar{n}_b))$, where $\bar{n}_b$ denotes the mean electron density at the cell-boundary. The equilibrium condition $p=0$ leads to

$$\bar{\mu}_{cell} = \varepsilon^{(0)}(\bar{n}_b). \tag{2.30}$$

In Fig. 2.5a $\varepsilon^{(0)}(\bar{n}_b)$ is compared with $\bar{\mu}_{cell}$ calculated by HEINE and HODGES /2.8/ as described above. For the alkalis and Be, Al, Zn and Mg the agreement is reasonably good. HODGES suggested that the deviations for the other elements may be due to nonlocal effects in the pseudopotentials on account of larger d cores.

For the transition metals NIEMINEN and HODGES /2.10/ expressed the kinetic energy in E_{cell} through (2.11) by wave functions. This leads to a relation between the chemical potential, and the wave functions and their derivatives at the cell boundary. The chemical potentials derived from this relation are shown in Fig. 2.5b. For a series of elements the agreement with the band structure estimates is satisfactory, but there are also large deviations.

Once $\bar{\mu}_{cell}$ is known the potential difference $\Delta\phi_{cell}$ can be obtained from (2.4) with the help of measured WF's. For nontransition metals $\Delta\phi_{cell}$ varies between -0.1 and 4.7 eV /2.8,9/, while for transition metals it lies between 2.6 and 6.0 eV /2.10/.

In the spherical cellular approximation $\Delta\phi_{cell}$ is related to a surface dipole layer which may be characterized in the following way. A reference electron distribution for the actual surface distribution is defined by continuing the bulk distribution up to the surface and cutting it off abruptly at the boundary of the surface cells. In Fig. 2.6 the deviation of the actual from the reference distribution is shown. It can be described as a dipole layer, and this dipole layer is responsible for the potential barrier $\Delta\phi_{cell}$.

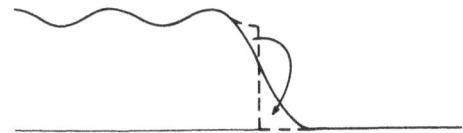

Fig. 2.6. Schematic plot of the electron density profile (full line) and the reference distribution described in the text (dotted line)

2.2.4 Uniform-Background Model and Its Extensions

In the uniform-background model the ion core charges are completely smeared out.
For the purpose of surface calculations the ion cores in a half-space are represen-
ted by a constant charge density which drops to zero at the surface plane (cf.
Fig. 2.7a), i. e.,

$$n_+(x) = \begin{cases} n_0 & , \ x \leq 0 \\ 0 & , \ x > 0 \end{cases} . \tag{2.31}$$

This assumption is particularly appropriate for metals with conduction electrons
arising from s- and p-shells. It is not suitable for d-electron metals,[8] semiconductors, or insulators.

In 1936 BARDEEN /2.4/ employed the uniform-background model in his pioneering
work on the WF of Na. He performed a partly self-consistent Hartree-Fock calculation.
More recently different forms for the surface potential were discussed /2.52-55/
until BENNET and DUKE /2.56,57/ performed a wave-mechanical calculation which is
self-consistent within the limits of an analytic expression for the electron density. Besides wave-mechanical treatments the statistical method based on (2.21)
has been used frequently /2.58-62/. The most detailed study of this kind is due to
SMITH /2.60,63/. Compared with the wave-mechanical method the statistical method
is much easier to handle, but the former is superior to the latter in so far as
the kinetic energy is much better approximated. Therefore we will discuss here
the most complete and exact wave-mechanical investigation which is due to LANG
and KOHN /2.11,12,64/.

a) Lang-Kohn Theory

LANG and KOHN /2.11,12,64/ adopted the density-functional formalism described in
Sect. 2.2.2 with the correlation energy expressed by Wigner's formula (2.17). The
equations in Sect. 2.2.2 are considerably simplified because of the one-dimensional
nature of the model.

The densities and potentials that result from the calculations of LANG and KOHN
for $r_s = 4$ are plotted in Fig. 2.7. The electrons protrude beyond the sharp edge of
the background charge density, thus setting up a dipole layer. This dipole layer
gives rise to the surface dipole barrier $\Delta\phi_{av}$ of the electrostatic potential that
is shown in Fig. 2.7b. The difference between the electrostatic potential ϕ and

---

[8] For a discussion of the applicability to d-electron metals see [Ref. 2.12, pp.
256, 263].

the effective potential v_{eff} is the exchange and correlation potential v_{xc}. The energies of the occupied electron states range from $v_{eff}(-\infty)$ to $\varepsilon_F = v_{eff}(-\infty) + \varepsilon_F^{(0)}$, where $\varepsilon_F^{(0)} = \hbar^2 k_F^2/2m$ denotes the Fermi energy relative to the conduction band bottom. The WF is given by $\phi_u = \phi(+\infty) - \varepsilon_F = \Delta\phi_{av} - \bar{\mu}_{av}$ with $\bar{\mu}_{av}$ taken from (2.28).

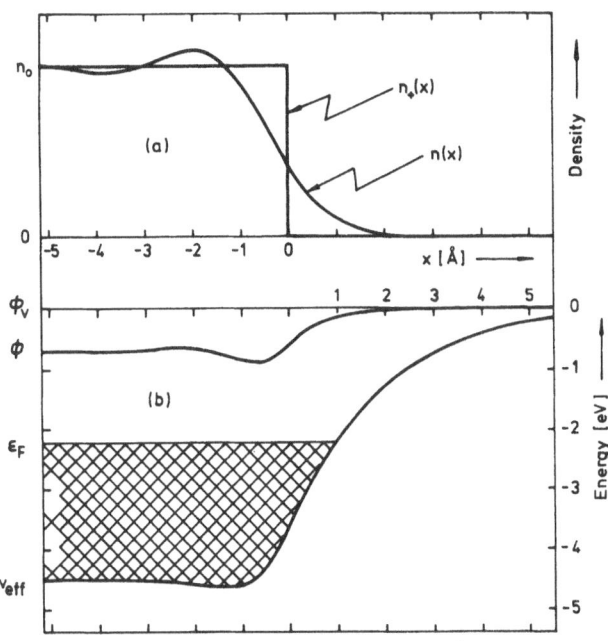

Fig. 2.7. Charge densities and potentials as computed by LANG and KOHN /2.11/ for the uniform-background model with $r_s=4$, plotted against the distance x normal to the surface. (a) background density, n_+, and electron density, n. (b) electrostatic potential, ϕ, effective potential, v_{eff}, and Fermi energy ε_F

Fig. 2.8 shows the energies and potentials relevant to the WF as functions of r_s. Note that the exchange and correlation contribution to the WF, μ_{xc}, is large in the whole metallic density range, as is to be expected from the qualitative discussion in Sect. 2.2.1. For $r_s=6$ almost the total surface barrier of the effective potential is due to exchange and correlation effects. The Fermi energy $\varepsilon_F^{(0)}$ and the dipole barrier $\Delta\phi_{av}$ change considerably from $r_s=2$ to $r_s=6$, whereas the WF ϕ_u varies by only 1.5 eV.

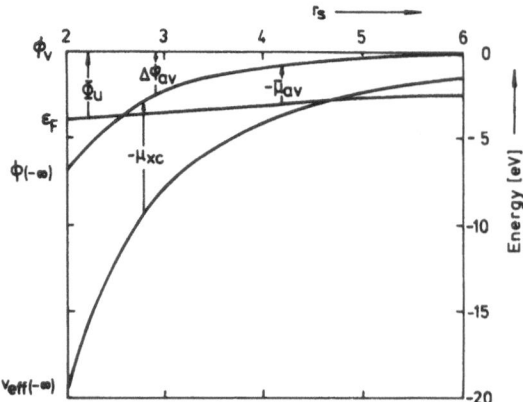

Fig. 2.8. Energies and potentials relevant to the WF, plotted against r_s. All energies are referred to the vacuum level ϕ_v. (Data from /2.11/)

A comparison between ϕ_u and WF's resulting from experiments on polycrystalline samples [Ref. 2.11, Fig. 2] reveals that, in view of the simple assumptions, the agreement between theory and experiment is very satisfactory for the alkalis and for Al, Pb, Zn and Mg, to which the uniform-background model can be expected to be best applicable. For the noble metals Cu, Ag and Au, the calculated WF's are 15-30% lower than the experimental values /2.11/.

The agreement with experiments found for simple metals could be improved further by perturbationally taking into account the ion-lattice /2.11/. The effect of the metal ions on the conduction electrons was represented by a pseudopotential of the form employed by ASHCROFT and LANGRETH /2.65/, $v_{ps}(r)=0$, for $r < r_c$, and $v_{ps}(r) = -Z/r$, for $r \leq r_c$. The ionic radius r_c had been determined to give a good description of the bulk properties. The difference between the total pseudopotential and the potential ϕ_+ due to the uniform-background charge was considered as a perturbation

$$\delta v_{ps}(\underline{r}) = \sum_\nu v_{ps}(|\underline{r}-\underline{R}_\nu|) - \phi_+(x). \qquad (2.32)$$

As shown by LANG and KOHN /2.46/, to the first order $\delta v_{ps}(\underline{r})$ has no influence on $\bar{\mu}_{av}$ but only on the dipole barrier $\Delta\phi_{av}$.

The corresponding corrections of the WF's depend on the orientation of the crystal plane, since δv_{ps} depends on the positions $\underline{R}_\nu$ of the atoms at the surface. In accordance with the arguments of STEINER and GYFTOPOULOS (Sect. 4.1) and of SMOLUCHOWSKI (Sect. 2.2.4d) the lowest calculated WF's are associated with the least densely packed faces.

The arithmetic average of the WF's computed for the (100), (110), and (111) faces of the cubic crystals, and results for the (0001) faces of the hcp crystals have been compared with WF's measured for polycrystalline surfaces [Ref. 2.11, Fig. 2]. For the simple s-p-band metals the agreement with experiments is improved to within 5-10%. In Table 2.1 for a selected set of metals the results of LANG and KOHN /2.11/ are compared with the predictions by STEINER and GYFTOPOULOS /2.2/ and with some new experimental results. For the simple s-p-bonded metals the results obtained by STEINER and GYFTOPOULOS and by LANG and KOHN agree very well with each other and with the experimental results, whereas for the noble and transition metals the predictions of STEINER and GYFTOPOULOS are in better agreement with the experiments.

b) Extensions of the Lang-Kohn Theory

After the fundamental work by LANG and KOHN many contributions concerning the uniform-background model have been published /2.66-77/.

Exact analytical results have been derived by BUDD and VANNIMENUS /2.66-68/. They determined the electrostatic potential at the surface (x=0) relative to its bulk value (at x=-∞) in terms of the mean energy per electron in a uniform electron gas of density n_0, $\varepsilon^{(0)}(n_0)$ [cf. (2.14-16)], and found /2.66/ $\phi(0)-\phi(-\infty)=n_0 d\varepsilon^{(0)}/dn|_{n=n_0}$. Combining this result with (2.4,14,25b,28) leads to the expression for the WF derived by MAHAN and SCHAICH /2.69/,

$$\Phi_u = \phi(+\infty) - \phi(0) - \varepsilon^{(0)}(n_0). \tag{2.33}$$

A further exact result has been found for the variation of the WF with the background density n_0. BUDD and VANNIMENUS /2.67/ showed that:

$$\frac{\delta\Phi_u}{\delta n_0} = 4\pi e^2 \int_{-\infty}^{+\infty} x \int_{-\infty}^{x} \delta n(x')dx'dx , \tag{2.34}$$

where $\delta n(x)$ is the normalized electron density induced at a uniform-background metal surface by a charge plane at infinity. This density has been extensively discussed by LANG and KOHN /2.70/ in a study of the image potential.

PEUCKERT /2.71/ used the Green's function method to determine the WF for the uniform-background model in the limit $r_s \to 0$. His result of 1.2 eV contrasts with 3.9 eV found by LANG and KOHN for $r_s=2$.

SAHNI et al. /2.72/ showed that the use of three different approximations for the correlation energy $\varepsilon_c^{(0)}$ leads to only small differences (of the order of 0.1-0.2 eV) between the resulting WF's.

Since recently the coefficient $\varepsilon_{xc}^{(2)}$ of the gradient terms (2.25c) has been calculated at metallic densities /2.32/, gradient terms can be included in the effective potential (2.24). It seems, however, that these terms have only a small influence, of typically 0.1 eV,[9] on the WF, and there is an indication that the inclusion of gradient terms can even worsen the calculated WF's /2.75/.

MONNIER and PERDEW /2.76/ have used a variational self-consistent method to
check the applicability of perturbation theory to the lattice potential $\delta v_{ps}(r)$ defined in (2.32). They solved the Schrödinger equation (2.23) for a parameterized
one-dimensional external potential and determined the parameters of this potential
by minimizing E_v [defined in (2.20)]. E_v was evaluated with the full three-dimensional external potential $\phi_+(x)+\delta v_{ps}(r)$. The one-dimensional electron density profiles calculated in this way are significantly different from the uniform-background
profiles. The dipole barriers $\Delta\phi$ for various crystal faces of the same element differ
by several eV.

Based on these electron density profiles MONNIER et al. /2.77/ calculated WF's
adopting the ΔSCF (change in self-consistent field) method of atomic physics. By this
method the WF can be written as the derivative of the surface energy with respect to
the surface charge density formed when electrons are removed from the metal. The
WF's calculated with this method differ from the results of LANG and KOHN /2.11/ by
typically 0.1 eV. The largest discrepancies are found for Al (see Table 2.1). For
the low index faces of Al the WF's calculated by MONNIER et al. are in better agreement with the most reliable experiment /2.78/ than the WF's found by LANG and KOHN
/2.11/. The calculations of MONNIER et al. do not show the anomalous trend $\Phi(100) >$
$\Phi(111)$ observed in the measurement of /2.78/ and in the theory of LANG and KOHN
/2.11/.

LANG and KOHN assumed that the lattice sites R_ν in (2.32) belong to a perfectly
periodic semi-infinite lattice. The experimentally observed compressions of the first
interlayer spacings of metals, however, vary up to 10% /2.79/. Several attempts have
been made to determine the lattice compressions theoretically /2.76,80-83/. The results seem to be not yet very reliable, except for the Al (111) face for which the
calculated relaxation shift is negligible /2.76/. MONNIER and PERDEW /2.76/ showed
that a shift of the first lattice plane by few percent of the bulk lattice spacing
can change the mean value of the pseudopotential δv_{ps} by as much as several eV,
but this change should be largly compensated by a corresponding change of the dipole
barrier $\Delta\phi$, so that the WF remains nearly constant.

---

[9] LAU and KOHN /2.73/ found gradient corrections between 0.3 and 0.7 eV. The corrections determined with more refined methods (ROSE et al. /2.74/), however, are
typically 0.1 eV.

c) Thin Metal Films

SCHULTE /2.84/ applied the uniform-background model to thin films by chosing a background charge density

$$n_+(x) = \begin{cases} n_0 & , \quad |x| \leq D/2 \\ 0 & , \quad |x| > D/2 \end{cases} , \qquad (2.35)$$

where D denotes the film thickness. The calculations were similar to those by LANG and KOHN /2.11/. If the thickness D is comparable with the de Broglie wavelength of the electrons confined in the film, the transverse motion of the electron is quantized. The energy spectrum splits into sub-bands. This so-called quantum size effect causes the WF to be strongly dependent on the film thickness.

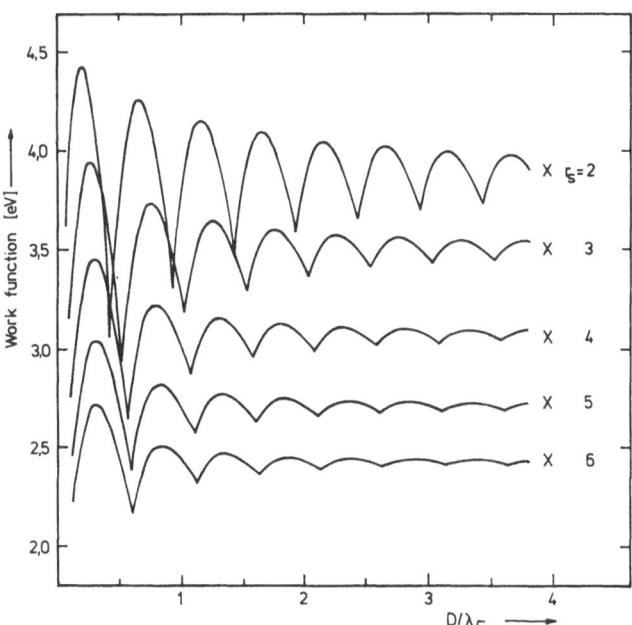

Fig. 2.9. WF of thin metal films plotted against D/λ_F for $r_s=2,3,\ldots,6$. D denotes the film thickness and λ_F the Fermi wavelength. The crosses on the right side of the curves correspond to the WF's calculated by LANG and KOHN /2.11/ for a model appropriate to the limit $D \to \infty$. (Based on /2.35/)

In Fig. 2.9 the WF's for background densities characterized by $r_s=2,3,...,6$ are plotted versus D/λ_F, where $\lambda_F=3.27r_s a_0$ denotes the Fermi wavelength. All curves show remarkable fluctuations, with cusps lying approximately $\Delta D=\lambda_F/2$ apart. The WF's calculated by LANG and KOHN /2.11/ are indicated by crosses on the right side of Fig. 2.9. The agreement with the WF's for the largest D considered is very satis-factory.

A comparison of the results shown in Fig. 2.9 with measured WF's of thin films is possible only under very restrictive conditions, which are discussed in the original paper /2.84/. From the experiments of JAKLEVIC et al. /2.85/, however, there remains very little doubt that quantized levels exist in real metal films. Therefore fluctu-ations of the WF as a function of film thickness should be observable. The first ex-perimental indication of these fluctuations has been provided by STARK and ZWICKNAGL /2.86/ who measured the field emission current from a glass tip covered by a thin Pd film. They found fluctuations of the emission current as a function of film thickness that may be related to the fluctuations shown in Fig. 2.9.

d) Qualitative Discussion of the Anisotropy of the Work Function

The empirical method of STEINER and GYFTOPOULOS discussed in Sect. 4.1 leads to the result that for a given material the larger (lower) WF's are associated with the more closely (loosely) packed surfaces. From a quite different point of view SMOLU-CHOWSKI /2.13/ had already reached that conclusion in 1941.[10]

SMOLUCHOWSKI studied a uniform-background model, which results from smearing out over a Wigner-Seitz cell the charge of each ion in a semi-infinite lattice. The boundary of this uniform-background charge is corrugated. The relatively smooth boundary shown in Fig. 2.10a represents schematically a closely packed surface whereas the rough boundary in Fig. 2.10b represents a loosely packed surface.

SMOLUCHOWSKI determined the electron density variationally employing the statis-tical approximation (2.13) for the functional E_v defined in (2.20), and a parameter-ized ansatz for the electron density at the corrugated surface. He included the VON WEIZSÄCKER /2.51/ gradient correction of the kinetic energy which is, compared with (2.19), too large by a factor of 9 and he neglected the total exchange and correlation energy ($\epsilon_{xc}^{(0)}$ and $\epsilon_{xc}^{(2)}$). Therefore his results cannot be expected to be quantitatively valid. Nevertheless they provide insight into the qualitative factors that determine the anisotropy of the WF, that is, the smoother the surface contours of the electron density, the lower is the contribution to E_v from the gradient term

---

[10] See also the approaches by LANG ([Ref. 2.12, pp. 268-270], discussion of the face dependence of the mean value of δv_{ps}) and by KELLY (/2.87/, corrugated, infinitely high potential barrier model).

of the kinetic energy, but at the same time the Coulomb energy rises. It is the interplay of these two energies that determines the electron density contours at the
surface and leads to contours which are considerably smoother than the background
boundary [Ref. 2.13, Fig. 2.10].

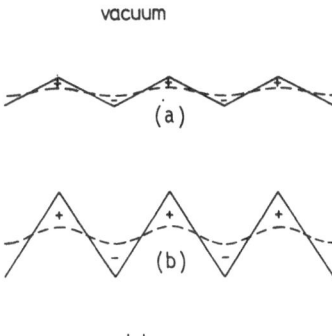

Fig. 2.10. Corrugated uniform-background model. The boundary of the background
charge (full line) and a contour of the electron density (dotted line) are plotted
for (a) a closely packed and (b) a loosely packed surface

This gives rise to a dipole moment oriented in a direction opposite to that of
the moment of a planar surface. This dipole moment is larger for the loosely packed
surface in Fig. 2.10a than for the closely packed surface in Fig. 2.10b. Since it
reduces the WF, the WF of a loosely packed surface is smaller than that of a closely
packed surface in agreement with the findings of STEINER and GYFTOPOULOS /2.2/, and
with the general trend of experiments (cf. Table 2.1) which are, however, mainly
performed for d-band metals to which the uniform-background model is not well suited.

The calculations of SMOLUCHOWSKI have been improved /2.87-89/, but the corrugated
uniform-background model seems to give no satisfactory quantitative results (see the
discussion in Sect. 10 of LANG's review /2.12/). It should be more appropriate to
include a pseudopotential perturbationally or in a variational self-consistent way
as discussed in the previous sections. In principle the best method is to perform
fully self-consistent wave-mechanical calculations which include the lattice potential from the beginning.

The few computations of this kind which are available at present will be discussed in the following section. In Fig. 2.11 a contour plot of the charge density
at a Cu surface is shown (from GAY et al. /2.90/). It shows clearly a smoothing

of the electron density, thereby substantiating the main feature of SMOLUCHOWSKI's model.

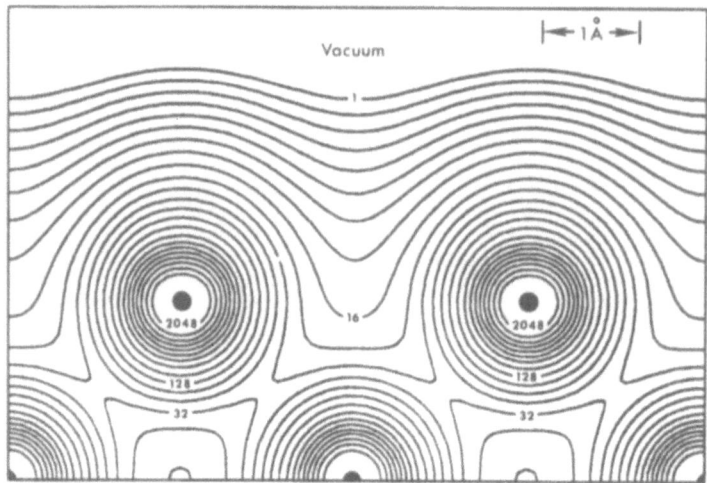

Electronic Charge Density at a Copper (100) Surface

<u>Fig. 2.11.</u> Contour plot of the electron density at a Cu (100) surface in a lattice plane perpendicular to the surface /2.90/

2.2.5 Wave-Mechanical Calculations for Lattice Potentials

Starting from lattice potentials, completely self-consistent wave-mechanical calculations have been performed for the three simple s-p-bonded metals Li (ALLDREDGE, KLEINMAN /2.91/), Na (APPELBAUM, HAMANN /2.92/), and Al (CHELIKOWSKY et al. /2.93/), for which the uniform-background model applies, too, and for the d-band metals Nb (LOUIE et al. /2.94/) and Cu (GAY et al. /2.90/). In the following we will give a short description of the main features of these calculations. The reader interested in further details is referred to the original papers and to the reviews by APPELBAUM and HAMANN /2.95/ and by SMITH /2.63/.

The models adopted range from a three layer film /2.90/ to a lattice filling a half-space /2.92/. The most recent calculation by GAY et al. /2.90/ has been done for a three layer Cu-film. As the authors pointed out, their result is possibly

incorrect due to the thin film geometry. The deviations due to the film geometry
can be estimated from the results for uniform-background films shown in Fig. 2.10.
For film thicknesses corresponding to a three layer Cu-film they amount to about
0.5 eV. In the calculations for Li, Al, and Nb films 13, 12, and 9 layers, respectively, have been adopted. Therefore the errors due to the film geometry are probably
negligible.

All calculations, except those for Cu, employ pseudopotentials for the ion cores.
Therefore the diagonalization of the Hamiltonian in a plane wave basis is numerically
feasible, and essentially this has been done for Li, Al, and Nb. For Na, APPELBAUM
and HAMANN /2.92/ used a mixed representation for the wave functions (plane waves
parallel to the surface and Fourier components depending on the coordinate perpendicular to the surface), determined band states with complex quasi-momenta perpendicular
to the surface, and used a matching procedure to find the total wave functions. For
Cu no pseudopotential was used and accordingly GAY et al. /2.90/ represented the
wave functions as linear combinations of atomic orbitals (ground state orbitals 1s
to 4s).

Exchange and correlation effects between the conduction electrons were in all
cases considered by a local exchange and correlation potential. $v_{xc}^{(0)}$ with Wigner's
expression (2.17) for the correlation energy was used for Li and Na, and the $x\alpha$-
potential (2.26) with $\alpha = 0.8$ for Al and Nb. For the Cu layer $v_{x\alpha}$ was used, too,
but with $\alpha = 2/3$, i.e., correlation effects were not included.

In Table 2.1 the theoretical results discussed in this section are compared with
some new experimental results taken from /2.96/. Compared with the experimental
values the results for the most sophisticated lattice potential models are considerably too large for Li, Al, and Cu, they are too low for Nb. The agreement for Na is
satisfactory. For Li, ALLDREDGE and KLEINMAN /2.91/ pointed out that the choice of
a pseudopotential parameter (E_{rep}) was not optimum with respect to the band structure and that for this reason their result is too large. The reason for the dis-
crepancy in the case of Al is not known. For Cu, as already discussed, the deviation may be partly due to the three layer geometry. Judging the result for Cu the
extraordinary neglect of all correlation effects should, however, also be taken
into account.

Table 2.1. Theoretical and experimental WF's for a selected set of metals. The predictions from the empirical treatment of STEINER and GYFTOPOULOS have been obtained
from (4.4,6) with R' and r_m taken from [Ref. 2.97, Tables 2-1 and 3-1], and with v_m
from [Ref. 2.98, Table 11-1]. For the fcc crystals the fractional bond numbers n''
have been neclected. The results of the Lang-Kohn theory are taken from [Ref. 2.11,
Table 2]. Two values for the same crystal face correspond to two different pseudopotential core radil r_c. The first entry corresponds to the r_c value which yields
agreement with experiment for a wider range of bulk properties. The experimental
values are taken from /2.96/. Values for polycrystalline samples or thin films without a defined surface orientation are marked with an asterisk

| Metal, Struc-ture | Face | Work Function [eV] | | | | Experiment |
| | | Theory | | | | |
		STEINER, GYFTOPOULOS /2.2/	LANG, KOHN /2.11/	MONNIER et al. /2.77/	Lattice Pot.	/2.96/
Li bcc	111	2.58	3.25 2.30			
	100	2.61	3.30 2.40		3.71 /2.91/	2.9
	110	2.75	3.55 2.40			
Na bcc	111	2.39	2.65	2.76		
	100	2.40	2.75	2.84	2.71 /2.92/	2.75
	110	2.52	3.10	3.13		
K bcc	111	2.24	2.35			
	100	2.25	2.40			2.30
	110	2.35	2.75			
Cs bcc	111	2.14	2.20 1.80			
	100	2.14	2.30 1.90			2.14
	110	2.23	2.60 2.25			
Al fcc	110	3.73	3.65	4.02		4.06
	100	3.92	4.20	4.25		4.41
	111	4.12	4.05	4.27	5.17 /2.93/	4.24
Cu fcc	110	4.65	3.55			4.48
	100	4.99	3.80		5.6 /2.90/	4.59
	111	5.32	3.90			4.98
Ag fcc	110	4.17	3.35			4.52
	100	4.45	3.55			4.64
	111	4.74	3.70			4.74
Au fcc	110	4.65	3.50			5.37
	100	4.99	3.65			5.47
	111	5.32	3.80			5.31
Nb bcc	111	3.94				4.36
	100	4.07			3.6 /2.94/	4.02
	110	4.75				4.87
Mo bcc	111	4.27				4.55
	100	4.44				4.53
	110	5.24				4.95
W bcc	111	4.47				4.47
	100	4.66				4.63
	110	5.52				5.25

2.3 Work Function Changes Induced by Adsorbates on Pure Metals

The theory of WF changes induced by adsorbates is to a large extent based on the classical model due to LANGMUIR /2.14/, and on the quantum-mechanical treatment by GURNEY /2.15/.

2.3.1 Classical Model

LANGMUIR /2.14/ explained the observed alkali-induced WF changes by an ionization of the adatoms. He assumed that the valence electron of an alkali adatom is transferred to the metal. The charge density of this additional metal electron is concentrated near the surface where it screens the field of the alkali ion.[11] The ion, together with the density of the screening charge, is characterized by a dipole moment p. The WF change $\Delta\Phi$ is proportional to the number of adatoms N_a per unit area, and is given by the Helmholtz equation[12]

$$\Delta\Phi = -4\pi ep N_a .\tag{2.36}$$

If p is assumed to be independent of N_a, (2.36) describes a linear change of the WF with N_a, in contrast to the observed behavior, which is discussed in Sect. 5.2. The deviation from linearity is attributed to a depolarization of assumed point dipoles by the Coulomb field of all the other point dipoles. Based on TOPPING's /2.99/ formula for this field the WF is given by

$$\Delta\Phi = - \frac{4\pi ep_0 N_a}{1+9\,\alpha\,N_a^{3/2}} ,\tag{2.37}$$

where p_0 is the initial dipole moment (in the limit $N_a \to 0$), and α denotes an effective polarizability. Many experiments have been analyzed with this formula, (cf., e. g., /2.100,101/). The physical origin of the depolarization is, however, not clear /2.101/.

---

[11] With the difference that the signs of the screened and of the screening charge are reversed, this corresponds roughly to the picture sketched in Fig. 2.4.

[12] In this review p is always identified with the dipole moment of the real charges, and not with the dipole p' formed by the charge of the adatom and its image. p' differs from p by a factor of 2 and correspondingly, if p' is used, the Helmholtz equation contains a factor 2 instead 4. If the field of the real dipole is idealized by the field of a point dipole, clearly p' has to be inserted as dipole strength.

2.3.2 Quantum-Mechanical Model

GURNEY /2.15/ revealed the character of the depolarization by a quantum-mechanical
treatment. He discussed the change of an atomic valence electron energy level when
the atom approaches a metal surface, cf. Fig. 2.12. At large separations the overlap between the metallic and the atomic wave function is very small, so that the
atomic wave function can be considered as an eigenstate with a well-defined energy
ε_a. If at smaller separations the wave functions begin to overlap, there is no
longer a well-defined atomic state. An electron on the adatom can tunnel to the
metal, and the atomic level is broadened. Its width 2Γ is related to the tunneling
time τ via $2\Gamma \approx h/\Gamma$. Besides the broadening, the mean energy ε_a' of the atomic resonance level is shifted. Both effects are shown schematically in Fig. 2.12. According
to GURNEY the depolarization is due to a coverage dependent downward shift of the
resonance level, which will be discussed in Sect. 2.3.4.

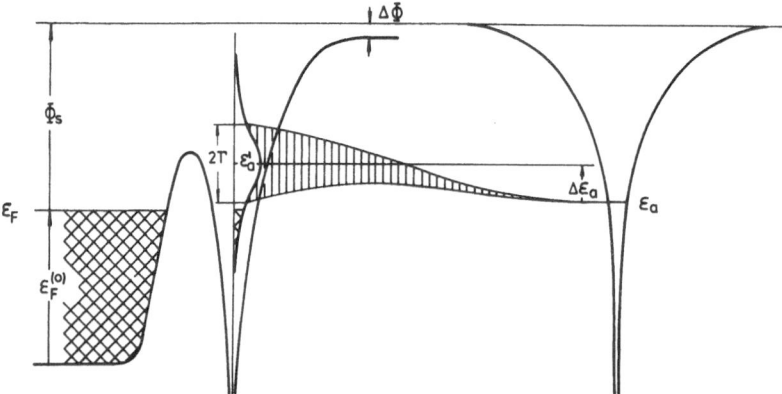

Fig. 2.12. Energy-level diagramm relevant to an alkali adatom at and near a metal
surface. See text

GURNEY's picture is the basis of several quantitative studies of the WF change
/2.102-105/. In the following an attempt is made to present an amalgamation of them
together with a critical view of some basic assumptions.

A common feature of all physical treatments[13] of the WF change $\Delta\Phi$ is the assumption that $\Delta\Phi$ is due exclusively to a change of the charge distribution, $\delta\rho(\underline{r})$, accompanying the adsorption. This assumption is justified by (2.4). It leads, via Poisson's equation, to the WF change

$$\Delta\Phi = -4\pi e \int_{-\infty}^{+\infty} x\delta\rho(x)dx, \tag{2.38}$$

with $\delta\rho(x) = \int_F \delta\rho(\underline{r})dydz/F$, where F denotes a large surface area.

Thus the problem is reduced to a calculation of the charge distribution $\delta\rho(\underline{r})$. If it is assumed that each adsorbed atom bears an effective charge q and that all the atoms have the same distance x_a from the surface, $\delta\rho(x)$ contains a δ-function with strength $N_a q$, and a part $\delta\rho_{sc}(x)$ from the screening charges. The WF change is then given by (2.36) with $p=(x_a-x_0)q$, where x_0 is the center of gravity of the screening charge density $\delta\rho_{sc}(x)$. (For an illustration of the relevant positions see Fig. 2.13). The only difference from LANGMUIR's model is that q is in general not equal to the charge of a singly ionized atom /2.106/.

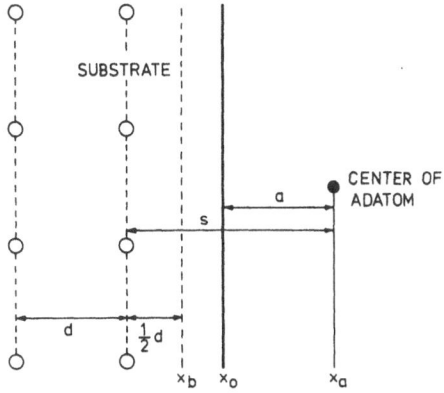

Fig. 2.13. Schematic plot of various positions and distances discussed in the text. Open circles represent positions of centers of substrate ions. (Based on /2.70/)

---

[13] In addition to $\Delta\Phi$, given by (2.38), GYFTOPOULOS and LEVINE /2.20/ considered an electronegativity barrier. The physical origin of this barrier is, however, not clear, cf. also /2.21/.

The methods of calculating the WF change differ in the way the distance $a = x_a - x_o$ and the charge q are determined.

GADZUK /2.102/ calculated the screening charge density and thereby the distance a with a linear response formalism similar to that used by LANGER and VOSKO /2.107/ for an impurity in a uniform electron gas. A more rigorous treatment is due to LANG and KOHN /2.108/, LANG /2.109/, and BUDD and VANNIMENUS /2.110/. They showed that within the limits of the linear response theory the response to a discrete point charge is the same as that to a charge sheet. Including the external potential due to this charge sheet in the calculations described in Sect. 2.2.4, they determined the distance a as a function of the distance between the charge sheet (at x_a) and the position x_b of the uniform-background surface (see Fig. 2.13). The result is shown in Fig. 2.14.

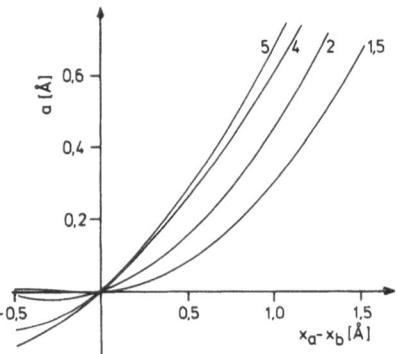

Fig. 2.14. Dipole length a as a function of the charge distance $x_a - x_b$ from the background surface. For the distances refer to Fig. 2.13. The curves correspond to $r_s = 1.5, 2, 4,$ and 5. (Based on /2.110/)

The actual distance $x_a - x_b$ can be derived from the separation s between the adatom and the upper atomic layer together with the interplanar spacing d (see Fig. 2.13). Since there are almost no measurements of s, GADZUK /2.102/ used a billiard ball geometry to determine s.

If the adatom is very close to the surface, nonlinear effects will be signifi- cant /2.111/, and the atomic "roughness" of the surface becomes of importance. Therefore these results are best suited to adatoms with large radii, like K, Rb, and Cs, on the most closely packed crystal faces ((110) for bcc, (111) for fcc, and (0001) for hcp lattices), for which the uniform-background model is most appro- priate.

From the above we can conclude that a determination of the distance a is still subject to considerable uncertainties. Therefore a has either been treated as an arbitrary parameter /2.112,113/, or it has been used to fit the measured WF changes /2.104,105/. HARTMAN /2.103/ fitted the distance s_A for one adsorbate A on a given crystal face to the measured initial dipole moment and to the binding energy. The distances s_B for adsorbates B on the same crystal face have been determined with the help of the ionic radii $r_{A(B)}$ of the adatoms A(B) from

$$s_B = r_B - (r_A - s_A). \qquad (2.39)$$

The effective charge q is determined by the occupied portion of the resonance level (see Fig. 2.12). It can easily be computed once the shape and the position of this level, i.e., the local density of states, are known. The local density of states is defined by

$$n_a(\varepsilon) = \sum_m |<a|m>|^2 \, \delta(\varepsilon - \varepsilon_m), \qquad (2.40)$$

where $|a>$ denotes the unperturbed atomic state and $\{\varepsilon_m, |m>\}$ characterize the eigenstates of the complete system of substrate plus adsorbate.

2.3.3 Newns-Anderson Formalism

To determine the local density of states extensive use has been made of the so-called Newns-Anderson formalism /2.16/.[14] In the following short discussion of this formalism we present it in its simplest form, i.e., we neglect all magnetic effects (due to the intra-atomic Coulomb interaction), and we consider only one atomic adsorbate state. The generalization to additional atomic states is straightforward /2.104,105/. Magnetic effects are believed to have no great influence on dipole moments of alkali adatoms on metals /2.114/.

For other adsorbates, like H, N, O, and CO, much more refined methods have been employed /2.115/, but unfortunately no systematic applications of these concepts to WF changes are known to the author /2.116/.

We start from a one-particle Hamiltonian H with a potential $V = V_{ma} + V_a$, where V_a denotes the unperturbed atomic potential. The potential V_{ma} includes the metal potential and the interactions between the adatoms, and will be discussed in Sect. 2.3.4. The Hamiltonian H is diagonalized in a basis $\{|\underline{k}>, |a>\}$, which is assumed to be orthogonal. $|\underline{k}>$ denotes the metallic states and $|a>$ the unpertubed atomic state.

---

[14] GADZUK /2.102/ did not explicitly use the Newns-Anderson model, but his results are consistent with their model.

It is assumed further that only the matrix elements $\varepsilon_{\underline{k}} = \langle\underline{k}|H|\underline{k}\rangle$, $\varepsilon_a'' = \langle a|H|a\rangle = \varepsilon_a + \langle a|V_{ma}|a\rangle$, $V_{a\underline{k}} = \langle a|H|\underline{k}\rangle$, and $V_{\underline{k}a} = \langle\underline{k}|H|a\rangle$ are nonvanishing. The local density of states can be expressed by the Green's function

$$G_{aa} = \langle a|(\varepsilon-i0-H)^{-1}|a\rangle = (\varepsilon-\varepsilon_a'' - \Lambda - i\Gamma)^{-1},$$
(2.41)

where

$$\Lambda = P \sum_{\underline{k}} \frac{|V_{a\underline{k}}|^2}{\varepsilon-\varepsilon_{\underline{k}}},$$
(2.42a)

and

$$\Gamma = \pi\sum_{\underline{k}} |V_{a\underline{k}}|^2(\varepsilon-\varepsilon_{\underline{k}}).$$
(2.42b)

P stands for the principal part. In terms of G_{aa} the local density of states is given by

$$n_a(\varepsilon) = \frac{1}{\pi} \text{Im } G_{aa} = \frac{1}{\pi} \frac{\Gamma}{(\varepsilon-\varepsilon_a'' - \Lambda)^2 + \Gamma^2}.$$
(2.43)

If it is assumed that Λ and Γ do not depend on the energy, $n_a(\varepsilon)$ has the form of a Lorentzian with half-width Γ centered on $\varepsilon_a' = \varepsilon_a'' + \Lambda$. This Lorentzian is shown schematically in Fig. 2.12.

In terms of the local density of states the effective charge on the adatom is given by

$$q = e(1 - \langle n_a\rangle), \text{ where } \langle n_a\rangle = \int_{-\infty}^{\varepsilon_F} n_a(\varepsilon)d\varepsilon.$$
(2.44)

Since the parameters ε_a' and Γ, which determine the local density of states, depend on q, the charge q has to be determined in a self-consistent way /2.112/.

2.3.4 Applications of the Newns-Anderson Formalism

Assuming a Lorentzian for the level shape, the charge q is determined by the half--width Γ and by the position of the shifted level ε_a'. The half-width Γ can be determined from (2.42b) /2.102/ or considered as a fitting parameter /2.104,105/.

a) Adatom Energy Level Shift

A first estimate of the shift $\Delta\varepsilon_a'(N_a=0)$ at zero coverage can be obtained as follows. The valence electron on the adatom induces its own image and feels therefore the potential $-e^2/4a$, with a as defined above (see Fig. 2.13). The electron also inter-acts, however, with the image of the ion core. Since the position of the latter is·

not changed when the electron moves, the corresponding potential is given by $e^2/2a$. According to this simple picture there should be a total upward shift $e^2/4a$.[15] GADZUK /2.102/ determined the energy shift $\Delta\varepsilon_a'(N_a=0)$ for alkali adsorbates by an explicit evaluation of the matrix element $<a|V_{ma}|a>$ of the potential $V_{ma}(\underline{r})$, which includes both of the just described potentials for an electron at an arbitrary position $\underline{r}$.[16]

For greater coverages the interaction of an electron on an adatom with all the other adatom charges and with their screening charges has to be included in V_{ma}. The energy level shift due to this interaction can be approximated by the Coulomb potential ϕ_a at an adatom caused by all the other adatom charges and their images. The first term in an expansion of ϕ_a in powers of N_a is[16] $\phi_a=-18eapN_a^{3/2}$. Therefore to a first-order approximation the energy level shift is given by

$$\Delta\varepsilon_a(N_a) = e^2/4a - 18eapN_a^{3/2}. \tag{2.45}$$

Apart from this "direct" Coulomb interaction, the "indirect" substrate-mediated interaction between the adatoms /2.118/ could be important, but it has not yet been included in an analysis of the WF change by the Newns-Anderson formalism.

According to (2.45) the adatom level is shifted downward with increasing coverage. This shift leads to a decrease of the charge q on the adatom and thereby to a de- crease of the dipole moment. This mechanism for the depolarization has already been discussed by GURNEY /2.15/. It is responsible for the deviations from linearity in the WF vs. coverage curves for alkali adatoms on metallic substrates.

b) Results and Discussion

Based on the Newns-Anderson model, GADZUK /2.102/ and HARTMAN /2.103/ calculated initial dipole moments p_0 for alkali atoms on several metal surfaces. The dipole moment p_0 is related to the initial slope of the measured WF vs. coverage curve,

---

[15] In this picture it has been assumed tacitly that the screening charge follows the electron instantaneously. If the screening charge is assumed to be static (simulated by the image charge $e<n_a>$), the energy shift is given by $e^2(1-<n_a>)/2a$. The limits of validity of the static and adiabatic approximations, respectively, have been discussed by HEWSON and NEWNS /2.117/. Their main result is that, if the surface plasmon period (the characteristic response time) is much shorter (longer) than the lifetime τ of the electron on the adatom, the adiabatic (static) approximation should be valid. For Cs on Re or W, e.g., the adiabatic approxima- tion seems to be justified, whereas for Ni or Cu substrates the conditions for its validity may not be satisfied so well.

[16] See also the discussion in Sect. 5.2. Higher order terms have been evaluated by MUSCAT and NEWNS /2.105/.

via $p_0 = -\Delta\Phi'(0)/4\pi e$, if it is assumed that at low coverages the adatoms are uniformly distributed on the surface. This assumption is believed to be fulfilled in general for alkali adsorbates.

In Table 2.2 the results of GADZUK and HARTMAN are compared with experimental values. In view of the strongly simplifying assumption the agreement between theory and experiment is quite satisfactory.

Table 2.2. Comparison of calculated with measured dipole moments for alkali atoms on various Ni, W, and Ta faces. The experimental values for Ni and W substrates are taken from /2.103/, the values for the Ta (110) substrates are from /2.102/

Ads.	Substr.	Initial Dipole Moment p_0 [Debye]		
		Theory		Experiment
		GADZUK /2.102/	HARTMAN /2.103/	
Cs	Ni 100		6.67	
	110	5.51	6.17	3.50
	111		7.04	
	W 100	6.29	5.97	9.05 7.95
	110	9.12	7.37	17 14.6
	111	4.52	5.69	8.70 6.1
	Ta 110	8.72		8.58
Rb	Ni 100		5.85	
	110		5.33	
	111		6.17	
	W 100		5.16	
	110		6.47	
	111		4.86	
K	Ni 100		5.27	
	110	3.61	4.74	2.65
	111		5.55	
	W 100	4.19	4.57	5.75
	110	6.77	5.84	7.85
	111	2.51	4.26	4.75
	Ta 110	6.31		6.40
Na	Ni 100	3.44	3.56	3.60
	110	2.07	2.85	1.60
	111	4.26	3.80	3.70
	W 100		2.60	1.75
	110		4.03	2.05 2.5
	111		2.1	1.50
	Ta 110	4.26		4.43
Li	Ni 100		2.5	
	110		2.0	
	111		2.90	
	W 100		1.4	
	110		3.18	5.3
	111			1.5

BENNET /2.113/ studied the low coverage regime. He took the "direct" Coulomb interactions between the adatoms into account, and also considered short-range interactions due to the overlap of the atomic orbitals for both monatomic and diatomic adsorbed species on the W (100) face. The main result is that the net charge per atom is approximately the same for both species.

MUSCAT and NEWNS /2.104,105/ considered the whole range of coverage up to a full monolayer. They realized that the energy separation between the s and p valence orbitals of Cs is only 1.44 eV. Accordingly they took two atomic states, $|ns\rangle$ and $|np_x\rangle$, into account. With increasing coverage the Coulomb interactions between the adatoms and their images push the energy of the adatom orbitals down as discussed in Sect. 2.3.4a. At the same time, however, the adatoms are polarized in the electric field of all the other adatoms and their images. Both the lowering of energy levels and the polarization of the atoms lead to a depolarization of the surface dipole layer. MUSCAT and NEWNS showed that both effects are comparable /2.105/.

They estimated the half-width Γ of the s and p levels (assumed equal), and the positions of these levels at zero coverage, by fitting the experimental WF vs. coverage curves. For Cs and Re (0001) the measured curve can be fitted only with positions of the s level very close to 1.8 eV above the Fermi level. The half-width should be between 0.7 and 2 eV, which implies that the s and p level are strongly hybridized.

To conclude we can say that Gurney's model provides a satisfactory explanation for the WF changes due to alkali adsorption on metal surfaces. Unfortunately a first-principles determination of the parameters a, ε_a', and Γ is very difficult. If these parameters are, however, adjusted so that the measured WF curves are matched, valuable information about these basic parameters can be derived.[17] Moreover Gurney's qualitative arguments proved to be extremely useful for an analysis of many WF measurements not only for alkali adsorbates but also for nonmetals and molecules.

2.3.5 Tight-Binding Approximation

In two very recent articles MORAN-LOPEZ and TEN BOSCH /2.17,18/ studied the chemisorption of alkali atoms on s-p-band substrates within a simple tight-binding approximation. In their model the adatoms are distributed on the top positions of the (100) surface of a simple cubic substrate. For details of the calculations the reader is referred to the original article. We only remark that in contrast to the studies reviewed in Sect. 2.3.4 the static approximation has been used for the image potential (see footnote 15 on p.35).

---

[17] For examples see also the discussion in Sect. 5.2.

For a random distribution of adatoms no minimum was found in the WF vs. coverage curves /2.17/. In a subsequent paper MORAN-LOPEZ and TEN BOSCH /2.18/ showed that ordering of the adatoms is accompanied by a decrease of the WF which leads to a minimum in the WF vs. coverage curves. When an ordered structure is formed the distance between the adatoms increases. The increase of the interatomic distance leads to a reduction of both the electrostatic repulsion and the overlap between the adatoms. As a result the energy level ε_a' is upward shifted and the level width 2Γ decreases. Both effects lead, via (2.44) to an increase of the ion charge, and thereby to a lowering of the WF.

2.3.6 Applications of the Density-Functional Formalism

The density-functional formalism, which has been discussed in Sect. 2.2.2, has also been applied to adsorption problems /2.119-126/. In applications of the density--functional formalism the ion cores of the substrate are replaced by a uniform--background charge. The ion cores (nuclei) of the adsorbate have been simulated by point charges (Sect. 2.3.6a), or by a uniform slab of background charge (Sect. 2.3.6b).

a) A Single Adatom

The first calculation dealing with a single adatom was performed by SMITH et al. /2.119/ for an H adatom. They employed the extended Thomas-Fermi equation (2.21) and used linear response theory for the interaction between the H nucleus and the uniform-background substrate. The equilibrium metal-adatom distance was determined by minimizing the total energy. The only parameter in this treatment is the substrate r_s (r_s=1.5 was chosen). The dipole moment p_0 was found to be very small (~ 0.03 Debye) in agreement with experiments. The same method has also been applied to all alkalis by KAHN and YING /2.120/, who employed a pseudopotential for the ion cores. KAHN and RASOLT /2.121/ included the gradient correction to the exchange and correlation energy $\varepsilon_{xc}^{(2)}(n(\underline{r}))|\nabla n(\underline{r})|^2$ [see (2.13,18)] in their analysis. Thereby the dipoles calculated by KAHN and YING increased considerably. The results of KAHN and RASOLT are listed in Table 2.3. In view of the linear response formalism employed the agreement with the dipole moments measured on the closest packed surfaces [e.g., W (110), Ta (111), and Ni (111), see Table 2.2] is surprisingly good.

A completely self-consistent wave-mechanical calculation has been carried out by LANG and WILLIAMS /2.122/. They considered the case of H, Li, and O. Figure 2.15 shows the electron distribution near a Li adatom /2.123/. We see that electrons are transferred from the vacuum side of the adatom towards the metal. Compared with the assumptions usually made in applications of the Newns-Anderson formalism (see Sects. 2.3.3,4), that the adatom/metal system can be treated as an effective point

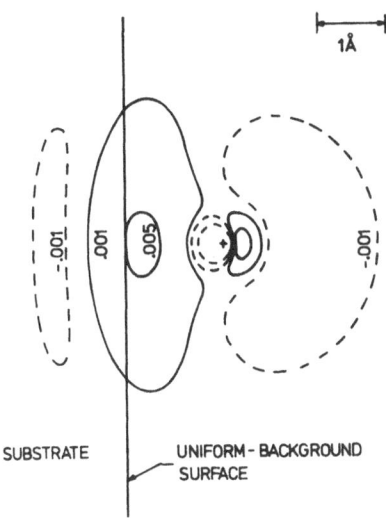

1Å

SUBSTRATE   UNIFORM-BACKGROUND
              SURFACE

Fig. 2.15. Contours of constant density in a (any) plane normal to the surface containing a Li adatom nucleus (indicated by +). Shown is the difference between the total electron density and a superposition of atomic and bare-metal electron densities [electrons/a_0^3]. (Based on /2.123/)

charge with a screening charge at the surface, this picture is much more complex. For the case of the smallest alkali, Li, it shows clearly a polarization of the adatom which was included in the study by MUSCAT and NEWNS (see Sect. 2.3.4 and /2.105,106/) only for the case of the largest alkali Cs.[18]

The dipole moments for H and Li are included in Table 2.3. The result for 0 is p_0=-1.7 Debye (r_s=2 was used for the substrate background charge). The agreement with measurements on transition metal surfaces is reasonable /2.122/, if it is taken into account that H and 0 cluster into islands even at very low coverages. For 0 on polycrystalline Al a negligible initial slope of the WF was measured. The disagreement with the dipole calculated for the uniform-background model has been explained by a penetration of 0 into the Al surface /2.124/.

b) Representation of the Adsorbate Layer by a Charge Slab

To describe the behavior of the WF for larger coverages LANG /2.125/ simulated the ions of the adsorbate layer by a uniform slab of background charge (see Fig. 2.16).

---

[18] See also the discussion of the dipole of an adsorbed Na atom in /2.126/.

The slab thickness d is assumed to be independent of coverage. It is determined by the interplanar spacing between the closest packed planes in the bulk alkali. The slab density n_a varies proportionally with the number of atoms per unit area N_a, that is, $n_a = N_a/d$.

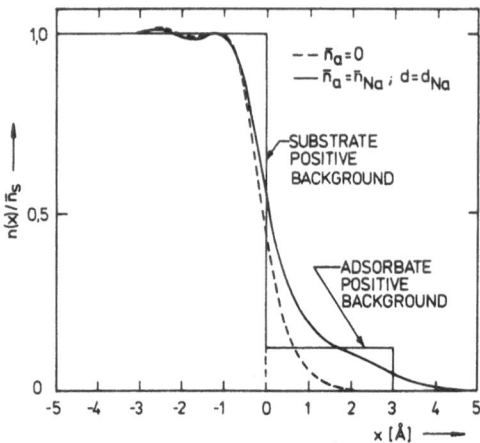

Fig. 2.16. Electron density distribution n(x) for a bare-substrate model ($r_s=2$) and for a model of a substrate with a full layer of adsorbed Na atoms. (Based on /2.125/)

LANG calculated the electron density wave-mechanically. A typical result is shown in Fig. 2.16. The deviation $\delta n(x)$ from the electron density of the bare-substrate model is shown in Fig. 2.17. Relative to the center of the adsorbate slab the center of gravity of $\delta n(x)$ is shifted to the left. Thus the uniform slab and the electron

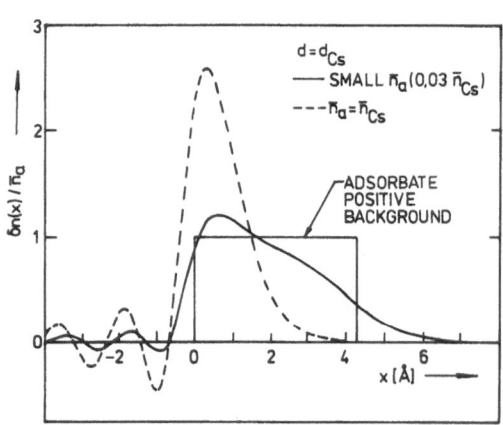

Fig. 2.17. Electron density change for low coverage, and full atomic layer coverage, of Cs. (Based on /2.125/)

charge density -eδn(x) form a dipole charge layer δρ(x) which, via (2.38), gives rise to a WF change. With increasing coverage the center of gravity of δn(x) moves to the right, and this corresponds to the depolarization already discussed in Sect. 2.3.4. The WF curves $\Phi(N_a)$ calculated by LANG show the typical behavior of measured WF curves (see Fig. 5.8).

Since the uniform-background model is not suitable for explanation of the WF's measured for bare d-band metal surfaces, LANG assumed a substrate density character- ized by $r_s=2$ and compared with measured data only those features which are relatively independent of the substrate WF, namely the WF minimum Φ_{min} (Fig. 2.18), the initial

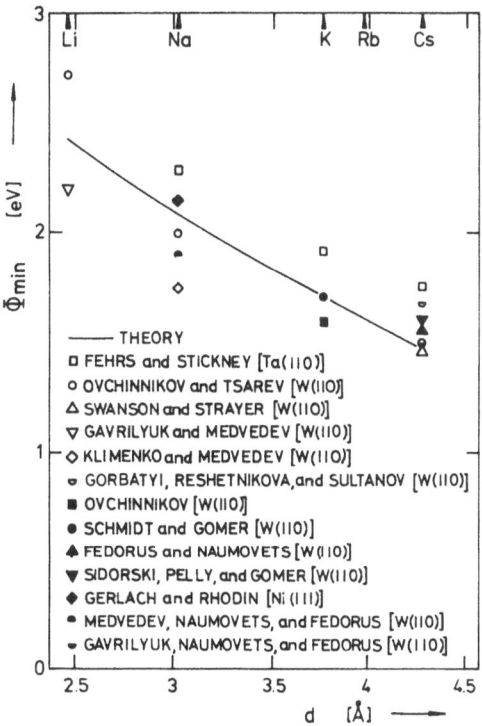

Fig. 2.18. Comparison between theoretical results derived by LANG /2.125/ and ex- perimental data for Φ_{min}. For the references to the experiments see /2.125/. (Based on /2.125/)

dipole moment p_0 (Table 2.3), and the WF $\Phi(\bar{N}_a)$ for a full monolayer of adsorbed atoms. The agreement is very satisfactory even for the initial dipole moment, though this model has not been designed to treat the case of very low coverages.

Table 2.3. Comparison of initial dipole moments calculated from the density-functional formalism. See text

	Initial Dipole Moment p_0 [Debye]		
Adsorbate	LANG, WILLIAMS /2.122/	KAHN, RASOLT /2.121/	LANG /2.125/
Cs		9.90	7.0
Rb		8.13	6.4
K		5.72	5.9
Na		3.86	4.2
Li	2.6	1.80	3.2
H	-0.5	0.06	

2.4 Work Function of Alloys

In their treatment of $Au_x Ag_{1-x}$ alloys GELATT and EHRENREICH /2.23/ published a very illuminating (though not exhaustive) contribution to the problem of WF's of alloys. Since this work seems to be suitable as starting point for further investigations, we will describe it in some detail.

An insight into the electronic structure of the alloy is provided by the following procedure. Imagine the pure Au and Ag crystals to be cut up into atomic cells, freeze the charge densities and stick the cells together to form the alloy. The atomic cells of Au and Ag are very similar in size so that they fill practically the whole space. The electron density of states in the frozen cells is the same as in the corresponding perfect crystals, namely $n(\varepsilon, x=1)$ for Au and $n(\varepsilon, x=0)$ for Ag. Alloying is accompanied by a charge transfer between the Au and Ag cells which leads to the density of states $n(\varepsilon, x)$. Once $n(\varepsilon, x)$ is known, the concentration dependent Fermi energy $\varepsilon_F(x)$, which can be compared with the measured concentration dependence of the WF, is easily obtained from

$$\int_{-\infty}^{\varepsilon_F(x)} n(\varepsilon, x) d\varepsilon = x N^{Au} + (1-x) N^{Ag} ,$$
(2.46)

where N^{Au} and N^{Ag} are the numbers of valence electrons per atom in the pure metals.

To calculate the density of states $n(\varepsilon, x)$ GELATT and EHRENREICH start from a model Hamiltonian H for the alloy system which contains as parameters the centers of gravity of the s- and d-bands, $\varepsilon_s^{Au(Ag)}$ and $\varepsilon_d^{Au(Ag)}$, and which allows, via hopping matrix elements, for nonzero bandwidths and, via hybridization parameters, for the hybridization of s- and d-states. All these parameters are fitted to reproduce the main features of the pure metal band structures.

The density of states $n(\varepsilon,x)$ is obtained via the configuration-averaged Green's function $G(z)=<(z-H)^{-1}>$, and calculated in the coherent potential approximation /2.127/

$$n(\varepsilon,x) = -(\pi N)^{-1} \text{Im Tr } G(\varepsilon+i0), \tag{2.47}$$

where N denotes the total number of atoms in the alloy.

A crucial problem is the placing of the energies of Au and Ag on a common energy scale. GELATT and EHRENREICH assumed that the difference between the Fermi energies ε_F^{Ag} and ε_F^{Au} is equal to the experimentally measured WF difference $\phi^{Au} - \phi^{Ag} = 0.9$ eV /2.128/. This assumption has not been justified by first-principles arguments. Indeed, as GELATT and EHRENREICH mention, there is no compelling reason to equate the vacuum levels of Au and Ag. The surface part $\Delta\phi$ of the WF should have no influence on the bulk alloy properties. Eliminating $\Delta\phi$ is, however, a very difficult task just because it is not a priori clear which reference level for the definition of $\Delta\phi$ (see Sect. 2.2.3) should be used.[19] The main motive for the adaption of the Fermi level difference to the WF difference by GELATT and EHRENREICH seems to be the agreement of their results with experimentally determined charge transfers, with optical absorption edges, and last but not least with the concentration dependence of the WF. GELATT and EHRENREICH point out that their procedure of choosing the Fermi energy difference may well not work in general.

An important feature of the calculations by GELATT and EHRENREICH is that they take into account the interplay between charge transfers $\delta n_j^{Au(Ag)}$ (j=s,d) and shifts of the energy parameters $\delta\varepsilon_j^{Au(Ag)}$. Because of the extended nature of the conduction electron wave functions GELATT and EHRENREICH neglected the s-band shifts $\delta\varepsilon_s^{Au(Ag)}$. The energy shifts of the more localized d-electrons are, however, strongly coupled to the charge transfers. They can be approximated by

$$\delta\varepsilon_d^{Au(Ag)} = U_{ds}^{Au(Ag)} \delta n_s^{Au(Ag)} + U_{dd}^{Au(Ag)} \delta n_d^{Au(Ag)}. \tag{2.48}$$

The constants $U_{ij}^{Au(Ag)}$ have been estimated to be of the order of 1 Rydberg /2.23/. The d-band shifts $\delta\varepsilon_d^{Au(Ag)}$ alter the Hamiltonian H and thereby also the charge transfers. This means that charge transfers and d-band shifts have to be calculated in a self-consistent way.

---

[19] HODGES and STOTT /2.37/ suggest that the chemical potential $\bar{\mu}_{cell}$ defined in Sect. 2.2.3 governs the charge transfer in simple s-p-band metals. The question as to whether this concept applies also to d-band metals has not yet been answered.

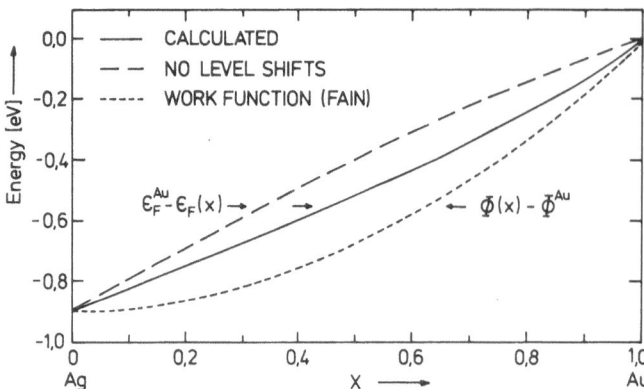

Fig. 2.19. Comparison of the calculated concentration dependence of the Fermi energy of Au_xAg_{1-x} alloys, with measured WF's. Note that the long-dashed curve calculated without allowing for d-band shifts bows the wrong way. (Based on /2.23/)

In Fig. 2.19 the calculated Fermi energy difference $\varepsilon_F^{Au} - \varepsilon_F(x)$ is compared with the WF difference $\Phi(x) - \phi^{Au}$ measured by FAIN and MCDAVID /2.128/, in one case neglecting the d-level shifts completely and in the other including it in a self-consistent way. Only in the latter case does the theoretical curve bow in the right direction.

In judging the results of GELATT and EHRENREICH it must be kept in mind that the surface of the alloy is not explicitly treated; the entire theory is designed for bulk alloy properties. Surface dependent quantities enter the theory only via the fit of $\varepsilon_F^{Ag} - \varepsilon_F^{Au}$ to the measured WF difference $\phi^{Au} - \phi^{Ag}$. To obtain more reliable results it is certainly necessary to treat the alloy surface more explicitly.

A first small step in this direction has been taken by DAVIDSON and FAIN /2.129/. They used the tight-binding method to obtain the ionization energies of Ag-Au and Cu-Au clusters with up to 135 atoms. The diagonal elements of their Hamiltonian are fitted to experimental valence state ionization potentials of the respective atoms, and the off-diagonal elements are given by the Mulliken-Wolfsberg-Helmholtz approximation /2.130,131/. A modification of the matrix elements at the surface is not considered. In certain cases the calculated ionization potentials show variations with the concentration similar to that of measured WF's. Although surface effects are considered to a certain extent the surface dipole potential $\Delta\phi$ is not explicitly included. To do this in their theory a modification of the orbitals at the surface atoms would be necessary /2.132/.

Further work in this challenging field is certainly highly desirable.

3. Experimental Procedures

J. Hölzl

3.1 Survey of Experimental Methods

In general, experiments to measure WF's can be divided into two groups: absolute
and relative. The first group is based on electron emission from a sample stimulated
either by high temperature (thermionic emission), by irradiation with light, (photoemission), or by the application of high electric field (field emission). The second
group (relative methods) basically takes into account the contact potential difference, which exists between the sample under examination and a reference electrode
when they are connected ohmically; accordingly this group can only produce WF data
that are relative to a certain (often unknown) WF, that of the reference electrode.

Since for the unique definition of the WF Φ an ideal monocrystal at $T = 0$ K, defined in infinite half-space in the absence of electric (and magnetic) fields, is
always assumed, the experimental conditions for the various types of measurements
must be discussed very carefully.

Clearly, both absolute and relative measurements of WF's will be strongly influenced by numerous parameters. For example, the finite size of the sample will give
rise to boundary "stray fields", while in the case of a polycrystalline sample,
patches will exhibit different WF values depending on whether high or weak applied
fields are used. Moreover, when using absolute methods, a lowering of WF $\delta\Phi$ is to
be expected as a consequence of external accelerating fields (Schottky effect).

3.2 Absolute Methods

As these methods are sketched here very briefly, the reader who is interested in
greater detail is referred to the excellent review articles by RIVIERE /3.46/ and
HAAS/THOMAS /3.47/.

3.2.1 Thermionic Emission

The density of saturation current J_0 produced by electrons emerging from a homogeneous metal surface at temperature T is given by the Richardson-Dushmann equation

$$J_0 = A(1-\bar{r}_e)T^2\exp(-\Phi/k_BT) \tag{3.1}$$

where the universal constant $A = 4\pi mk_B^2e/h^3 = 120$ Amp. cm^{-2} degree^{-2} with m and e
denoting the mass and the charge of the electron, respectively; h, Planck's and

k_B Boltzmann's constant and T the absolute temperature. Finally $\bar{r}_e$ is the mean value of the zero-field reflection coefficient for the incident electrons.[1] Since in all cases measurements of the saturation current involves application of an accelerating field, a correction must be introduced resulting from the lowering of the potential barrier at the surface, as shown in Fig. 3.1 (Schottky effect).

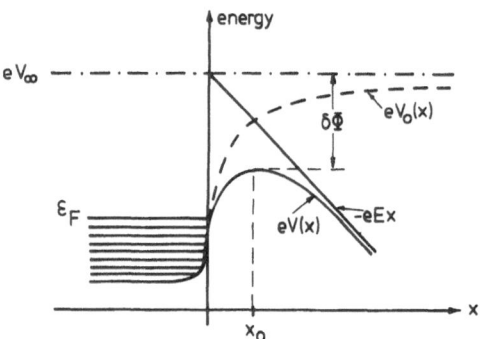

Fig. 3.1. Variation of potential energy near a metal surface:

$V_0(x)$ image potential

$V(x)$ potential in the case of an applied field of strength E

x_0 is the position of the maximum in the potential (barrier)

eV is the energy of the vacuum level for $x = \infty$

ε_F: Fermi energy

$\delta\Phi$: WF lowering due to the applied external field

Assumption of an image potential of the form $V_0(x) = e/16\pi\varepsilon_0 x$ leads to lowering of the potential barrier in the vicinity of the metal surface by an amount $\delta\Phi/e$. [ε_0: dielectric constant of the vacuum]. The WF variation $\delta\Phi$ vs. accelerating field E is given by

$$\delta\Phi = e \left[(e/4\pi\varepsilon_0)^{1/2} E^{1/2}\right]. \tag{3.2}$$

---

[1] The energy dependence of the electron reflection coefficient has been studied earlier by NIEDERMAYER and HÖLZL /3.11a/. Results obtained by these authors on W (110) have been evaluated with the use of a thermionic experiment done by HUTSON /3.11b/.

The position x_0 of the maximum in the potential barrier can be expressed as

$$x_0 = (e/16\pi\epsilon_0)^{1/2} E^{1/2}. \tag{3.3}$$

To determine the WF of the emitter, one must consequently measure the dependence of the saturation current J on temperature T and on applied field strength E. From so-called Schottky plots (ln J vs. $E^{1/2}$) one gets $J_0(T)$ by extrapolation to E=0.

The slope of the plot of ln (J_0/T^2) vs. $1/T$ (Richardson plot) leads to the so--called Richardson WF $\Phi*$

$$\Phi* = - k_B [d/d(1/T)] \ln (J_0/T^2). \tag{3.4}$$

Inserting (3.1) into (3.4) we obtain

$$\Phi* = \Phi - T(d\Phi/dT) - [k_B T^2/(1-\bar{r}_e)](d\bar{r}_e/dT) \tag{3.5}$$

showing that $\Phi*$ would only be the "true WF" Φ, if the temperature dependence of Φ and $\bar{r}_e$ vanished. $(d\bar{r}_e/dT)$ is commonly taken to be zero, but this is not the case for $(d\Phi/dT)$.

Thus the true WF $\Phi(T)$ is related to the Richardson (or "apparent") WF $\Phi*$ by

$$\Phi(T) = \Phi* + T (d\Phi/dT). \tag{3.6}$$

Note that $\Phi*$ is not the "true WF" for T=0 as might be expected from the above formula, because this would only be true if $(d\Phi/dT)$ were constant with T, which is not so.

When studying the thermionic emission from polycrystalline surfaces "patch--fields" should also be considered. The effect has been discussed by HERRING and NICHOLS /3.2/ for two limiting cases:

a) In case of a strong external field (x_0<<patch dimensions) each patch emits according to its individual work function. The measured work function of the whole surface is governed mainly by patches having low work function values since these show the most intensive emission.

b) In the case of a weak external field (x_0>>patch dimensions) all those patches for which $\Phi_i \le \bar{\Phi}$ emit as if they were a large single patch of work function $\Phi = \sum_i f_i \Phi_i$. The electrons emitted from these patches must overcome a potential barrier of height $\bar{\Phi}$. The patches with $\Phi_i > \bar{\Phi}$ emit according to their individual work functions. The work functions measured in this case would be close to $\bar{\Phi}$.

48

Two well-known experimental facilities for the determination of the work function by thermionic emission are the planar diode and the cylindrical diode methods as shown in Figs. 3.2,3.

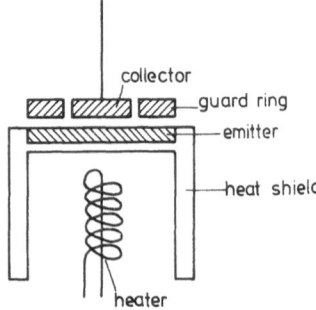

collector
guard ring
emitter
heat shield
heater

Fig. 3.2. Schematic drawing of the planar diode for WF measurement by thermionic emission

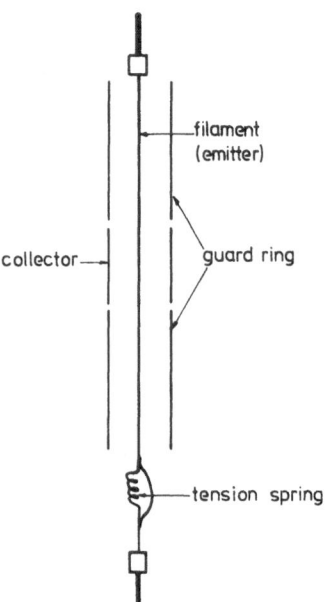

filament (emitter)
collector
guard ring
tension spring

Fig. 3.3. Schematic drawing of the cylindrical diode for WF measurement by thermionic emission

3.2.2 Photoelectric Method

The saturation photocurrent J_o coming from a uniform conductor with a work function Φ at temperature T with incident light of energy $h\nu$ is given by the FOWLER /3.3/ expression

$$J_o = B \cdot (k_B T)^2 f([h\nu - h\nu_o]/k_B T), \tag{3.7}$$

where $h\nu_o = \Phi$. In this expression B can be taken as constant close to the photoelectric threshold. f is a universal function. To determine $\Phi = h\nu_o$, J_o is measured either as function of ν (T = constant, FOWLER method /3.3/) or as a function of T (ν = constant, DU BRIDGE method /3.4/). Also in the photoelectric method a correction is necessary because of the applied electric field (photoelectric Schottky effect /3.4-7/).

Just as in the case of thermionic emission the relationship between the work functions of polycrystalline and of single crystal targets is considerably complicated. The weighting of the patches having different work functions corresponds to the weighting in the case of thermionic emission. This means that in the presence of strong accelerating fields the emission is largely determined by the patches with low WF, while in the case of weak accelerating fields one measures only the value of $\Phi = \Sigma_i f_i \Phi_i$. A plan view of an experimental arrangement used recently by BERGE et al. /3.1/ is shown in Fig. 3.4.

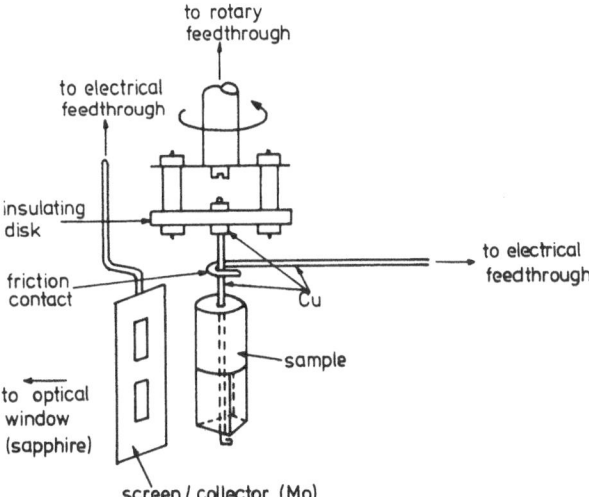

Fig. 3.4. Arrangement for photoelectric WF measurement as used by BERGE et al. /3.1/

3.2.3 Field Emission

If a strong accelerating field is applied to a metal surface a measurable electron
current is obtained. This is because the tunneling probability for electrons is increased, since the potential barrier at the surface is becoming smaller. The theoretical treatment of the effect follows the FOWLER-NORDHEIM equation /3.8,9/

$$J = \frac{e^3 E^2}{8\pi h \phi t^2(y)} \exp\left[\frac{-4(2m)^{1/2}}{3e} \frac{\phi^{3/2}}{E} \delta(y)\right], \tag{3.8}$$

where $t(y)$ and $\delta(y)$ are functions of a quantity y given by

$$y = (e^3 E)^{1/2}/\phi.$$

The determination of the WF is made by measuring the emission current J as function
of accelerating voltage $U(U = \beta E; \beta = $ geometrical factor) and temperature T. The
evaluation of the results is done according to the methods of reference /3.9,10/.

As the tip of the probe in such a field emission experiment is composed of
crystal surfaces of different orientations the measurement of a total emission
current determines only an average WF. This average is heavily weighted towards
those surfaces having lower WF. The "probe hole" /3.11,12/ technique enables one to
measure the emission current from any individual area of the cathode. This current
is thus from a particular crystal surface on the probe and hence the method allows
us to measure systematically all the individual WF's involved, and not simply the
averaged value.

Suitable data analysis combined with the help of this technique allows for example the observation of relatively rapid WF changes as adsorption on individual

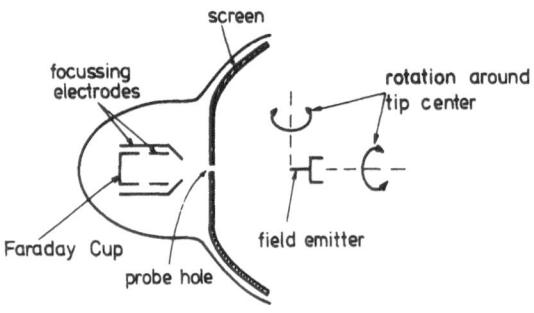

Fig. 3.5. Experimental arrangement for determination of WF by the field emission
probe hole technique as used by MALY /3.13/

crystal surfaces occurs. Such a procedure is presented by MALY /3.13/. The apparatus
used in this case is sketched in Fig. 3.5.

Recent work shows that for an exact determination of the WF using field emission
techniques, the emission current information must be supplemented by a study of the
distribution of the electron energies in order to account for band structure effects
/3.14/. Figure 3.6 shows a relevant apparatus which has been used by KUYATT and
PLUMMER /3.15/.

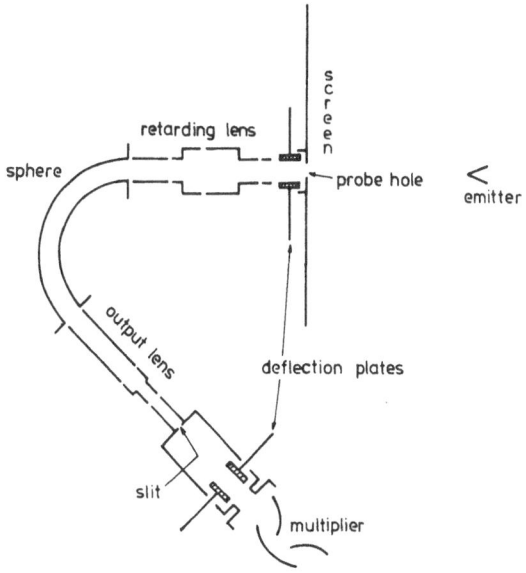

Fig. 3.6. Sketch of a field emission probe hole apparatus with energy analyser for
accurate WF determination as used by KUYATT and PLUMMER /3.15/

3.3 Relative Methods

The aim of the measurements indicated so far has been the determination of the absolute value of the sample work function. To this end differing electron emission
processes are used. The precision of the absolute methods of work function determination is, however, limited. The limitations arise from the approximations which
have to be made in the theoretical descriptions of the relevant physical processes,
and from the divergence between experiment and theoretical prediction. The relative
methods which we now describe are used mainly to observe temporal changes in the
work function of a metal surface. In very few cases only can one determine the
absolute value of the work function by measuring differences, as will be mentioned

below. All relative methods use the fact that between two metal samples with an external connection there exists a Contact Potential Difference (CPD), which is equal to the work function difference divided by the elementary charge. When the work function of one of the surfaces is altered as a result of some physical process, then the CPD, and hence the electric field between the two surfaces, is likewise altered. This charge resulting from the CPD variation can be compensated by applying an equal and opposite voltage in the circuit between the electrodes to its original value. The compensation of the field to obtain the original conditions is observed by different physical effects which react to changes in the field. Principally one uses either the anode current of a diode, used in the regime of low current and voltage (diode methods), or the displacement current in the connecting leads of the electrodes, created by an artificial change in the capacitance between the two electrodes (condenser methods). The precision with which the work function change can be measured by these methods depends mainly on the accuracy of the detection instrumentation. Work functions can be so determined with an accuracy from 0.01 to 10 meV. Since the relative CPD methods fall in two groups, the diode methods and the condenser methods, we shall describe them separately. The diode methods and their development for the measurement of surface potentials were excellently reviewed by KNAPP /3.16/. In the subsequent review we follow his presentation closely.

3.3.1 Diode Methods and Examples of Practical Configuration

In the diode method the work function is measured at the anode. Here we use the fact that when the work function Φ of the anode varies there is a parallel displacement of the characteristic curve of the diode. Quantitatively this displacement ΔU_a equals $\Delta\Phi_a/e$. This situation is sketched in Fig. 3.7.

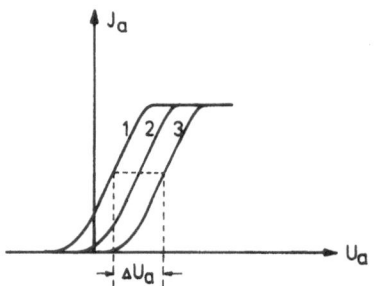

Fig. 3.7. Displacement of the J_a/U_a characteristic of a diode by change of anode WF: $\Phi_1 < \Phi_2 < \Phi_3$; $\Delta U_a = (\Phi_3 - \Phi_1)/e$

In practice the work function change is obtained not by observing the whole characteristic curve, but rather by maintaining anode current at a constant value by
readjustment of the anode potential. This necessary anode potential change is then
equal to the change in the CPD. Using a suitable electronic feedback system one can
plot automatically CPD versus time variation. According to where one works on the
characteristic line of the diode one can differentiate between two cases: the space
charge limited diode and the retarding field diode. The variation of potential

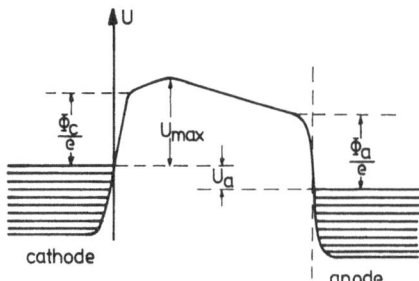

Fig. 3.8. Potential variation between cathode and anode in the space charge limited
diode

between cathode and anode in the case of the space charge limited diode is sketched
in Fig. 3.8.

The value of the anode current in this case is largely determined by the height
of the potential barrier (U_{max}) between cathode and anode. U_{max} is given by the
geometry of the apparatus, and the amount of space charge in front of the cathode
Φ_0, Φ_a and U_a. Using a simplified derivation one obtains the anode current J_0 versus
U_a as follows

$$J_a = B_g \, (\Phi_c + \Phi_a + eU_a)^n,$$ (3.9)

where B_g = constant (dependent on the geometry) and where $n \approx 1,5$. For high cathode
temperatures (which is usually the case) Φ_c is constant during the measurement and
therefore J_a depends only on the sum $(eU_a + \Phi_a)$. If one keeps J_a constant during
the measurement then the necessary change in U_a is equal to the CPD variation in
question.

The variation of potential between cathode and anode of the retarding field
diode is shown in Fig. 3.9. There exists a weak retarding field between cathode
and anode which the electrons, having kinetic energy, can overcome. The anode

54

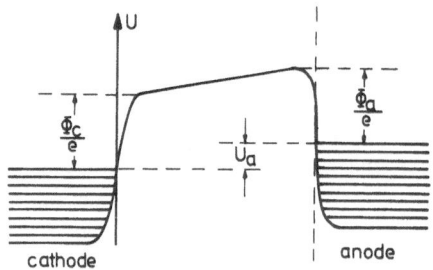

Fig. 3.9. Potential variation between cathode and anode in the retarding field
diode

current J_a in such a case can be calculated in an analogous way to the saturation
current in the case of a cathode in thermionic emission having a work function
$(eU_a - \Phi_a)$. The result is

$$J_a = A(1-\bar{r})T^2 \exp [(eU_a - \Phi_a)/k_B T], \qquad (3.10)$$

where A has the same meaning as in (3.1) and $\bar{r}$ represents the average reflection
coefficient at the anode. In this process the work function of the cathode has
basically no effect on the anode current, since the height of the potential barrier
is not related to the cathode work function. However, a more refined treatment of
this topic shows that there is in fact a small effect from Φ on the anode current
/3.17/. Additional consideration is necessary when the anode is polycrystalline or
otherwise patchy (e.g., has a surface coating). This consideration leads us to the
following main result: if the geometry of the experiment is so arranged that the
field strength in front of the anode is small compared with the patch fields (which
exist because of different WF's of the individual patches) then the diode method
produces a mean work function $\bar{\Phi} = \sum_i f_i \Phi_i$. Here f_i represents the fraction of surface
which has the work function Φ_i. On the other hand when field strength in front of
the anode is large compared with the patch fields we obtain a mean work function
$\bar{\Phi}$ which is heavily weighted towards those patches having lower work functions. Where,
however, the electron beam can be focused strongly to a diameter less than that of
patch size, the work functions of individual patches can be determined.

Practical Diode Configurations

Different diode configurations have been developed according to the different experimental questions to be answered. They are principally the following:

1) Spherical or cylindrical diodes
2) Crossed filament diodes
3) Electron beam facility
4) Scanning beam diodes

The spherical and cylindrical diodes (class 1) are used mainly to study the effects of gas adsorption on the work function of a metal film, evaporated on the walls of a glass sphere. Typical diodes of this type have been described by PRITCHARD /3.18/ and MIGNOLET /3.19/. Such a device is shown schematically in Fig. 3.10. The crossed

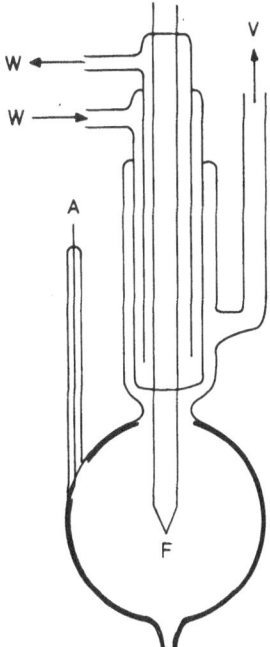

Fig. 3.10. Spherical diode as described by /3.18/
F: filament, A: anode contact, V: connection to vacuum system, W: thermostated water supply

filament diode (class. 2) is very simple in its construction and has been used, e.g., by HAYES /3.17/. Here the anode and the cathode consist of wires or ribbons arranged perpendicular to each other. The roles of anode and cathode are inter-changeable. This method is well suited to the examination of refractory metals, which can be cleaned easily by flashing processes. With this arrangement one can simultaneously do flash-desorption experiments. The mutual heating of the wires, however, imposes a lower limit on the temperature range of the method. In contrast to the methods listed under classes 1 and 2 the electron beam method has recently been highly developed with many variations. The classical configuration was described by ANDERSON /3.21/ in one of the earliest papers and is shown in Fig. 3.11.

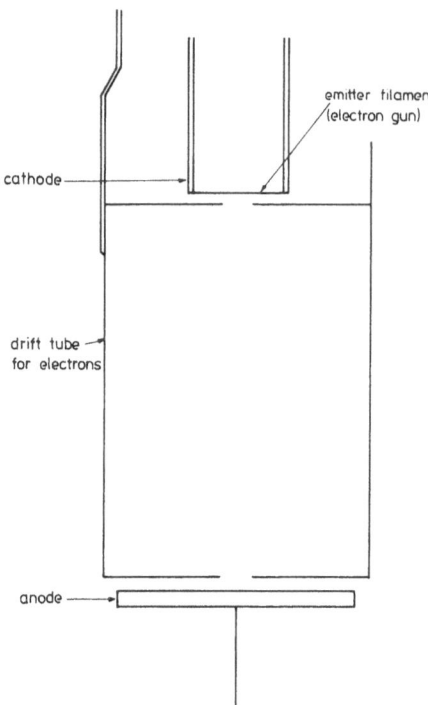

emitter filament
(electron gun)

cathode

drift tube
for electrons

anode

Fig. 3.11. Electron beam diode

The electrons emitted from a cathode are accelerated in a lens system (simply an electron gun) and are directed as a fine beam perpendicular to the anode. This type of diode is especially suited because of its geometry to the study of adsorption phenomena on small monocrystalline targets.

NATHAN and HOPKINS /3.20/ have presented a variation of the method. In this arrangement they used the electron source of a commercial Low Energy Electron Diffraction (LEED) system and measured continuously the work function variation with an accuracy of ± 1 meV.

A somewhat different method of measuring work functions of diodes is the so-
-called break point method. Here one uses the fact that theoretically there is a
sharp knee in the J_a/U_a characteristic of a diode when passing from the retarding
field mode to the saturated emission mode (when no space charge effects are assumed).
This knee corresponds to the point at which the applied anode voltage U_a equals
the CPD between anode and cathode. When the work function of the target (anode) is
varied then the knee is displaced accordingly.

An appropriate arrangement has been described by SHELTON /3.22/. In order to obtain a clearly defined knee SHELTON employed a thin, collimated electron beam which
was directed perpendicularly onto the target (anode) by an axial magnetic field.
In general, since the knee is rounded off by various effects (partly instrumental)
HERRING and NICHOLS /3.2/ suggested earlier using the crossing point of the two
limiting slopes.

The methods described up to now have given an averaged value of the work function
when using polycrystalline samples. However, HAAS and THOMAS /3.23,24/ have opened
up to the possibility of spatial resolution on the micron scale with their "beam
scanning" diode to be described below (class 4). The arrangement, Fig. 3.12, consists

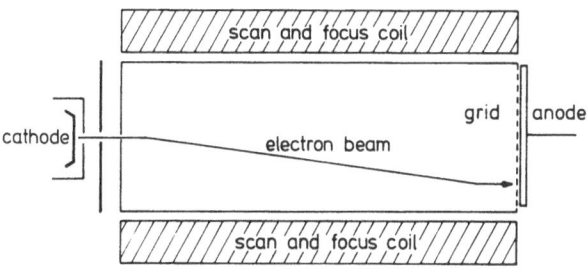

Fig. 3.12. Beam scanning diode as described by /3.24/

mainly of an electron gun producing a beam of diameter a few μm. The beam is always
directed perpendicularly onto the target and is scanned magnetically over the whole
target. In front of the target lies a fine grid which is biased to the same potential
as the final anode of the electron gun. This results in a high electric field of
the order of 5000 V/cm which is considerably greater than that of the patch fields.
Hence the work function of individual patches can be measured providing the beam
diameter remains smaller than the patches.

With the aid of the magnetic scanning system temporal changes in the work function of an arbitrary spot on the target can be observed. With the use of such a

scanning procedure only a short time is required for the electron beam to probe the entire surface of a target. The current impinging on the surface varies according to the locally changing work function. Thus, the use of suitable electronic detection enables a monitor to construct a picture of the work function's spatial distribution. This method is very well suited to the investigation of adsorption and diffusion processes on various surfaces. All the diode methods described above have the fact in common that they use electrons which are exclusively thermally emitted. On the other hand, HOLSCHER /3.25/ has described a diode set-up using a field emission cathode. Because the energy of such electrons is sharply defined it is possible to determine the work function of a target (=anode) absolutely. (In contrast to the case of using thermally emitted electrons where this is not possible).

Since the main uses of the diode method are in gas adsorption experiments, we make the following remarks about the limitation of its applicability. The gas pressure in an experiment must be less than 10^{-3} torr because of the effect on the electron mean free path. Because of possible chemical reactions with the hot cathode many gases cannot be used (e.g., oxygen, hydrogen, and numerous organic compounds). Possible undesirable effects of the electron beam on the adsorbate/substrate system such as are mentioned repeatedly in the literature can be dismissed as follows: even in the most unfavorable case the electron beam density would be only of the order of one electron per 30 surface atoms; furthermore, the kinetic energy of the electrons is mostly less than 1 eV. Thus, the disturbing influence of such electrons is very small when compared with that of the primary beam in Auger Electron Spectroscopy (AES) experiments.

3.3.2 Condenser Methods

General

As opposed to methods mentioned previously, the condenser methods use neither emission processes nor collected particles (electrons, photons) in order to determine the CPD between two metal surfaces, but rely on the fact that between two conducting connected metal surfaces there exists in general a potential difference (CPD) on account of the difference in work functions. The electrical field arising from this is used for measurement. Hence, there is no risk of changing the surface condition of the sample by the measuring procedure itself.

The range of application of these methods is far wider than for the others, with respect to the materials to be examined, the gas pressures in adsorption experiments and the temperature range.

a) Vibrating Capacitor Methods

Most condenser methods are based on the principle of the vibrating condenser given by KELVIN /3.26/ and by ZISMAN /3.27/. The apparatus comprises a condenser of variable capacitance, which is made up from the metal surface (target) to be examined and the reference electrode. In the external circuit, the two electrodes are connected via an ammeter and a variable voltage source. This arrangement is shown in principle in Fig. 3.13a,b,c together with the potential relationship between the two condenser plates.

If the capacitance C between the plates is changed in any way, then the current i, with the magnitude

$$i = U \, dC/dt \tag{3.11}$$

flows through the circuit, where U is the potential difference between the two condenser surfaces. U is given by the sum of the CPD ($\Delta\Phi/e$) between the target and reference electrodes and the voltage U_C of the variable voltage source.

In order to determine the CPD, the voltage U must be so adjusted that no current flows even, when the capacitance is changed. Then

$$U = \Delta\Phi/e + U_C = 0 \tag{3.12}$$

and

$$U_C = - \Delta\Phi/e. \tag{3.12a}$$

Changes in capacitance are usually produced by periodic vibration of one of the condenser electrodes around its equilibrium position. The condenser itself is often made of two plates of the same size, but depending on the experimental requirements, could be made of one plate and a much smaller reference electrode, (e.g., for the measurement of the CPD variation along one coordinate) /3.28,29/, or of two spheres /3.30/.

The excitation of the vibrating electrode can be performed mechanically /3.31/, electromechanically /3.30/, or electrostatically /3.32/. Other means of capacitance change are also possible, for example the rotation method of KOLM /3.33/ and later MITCHINSON et al. /3.34/, or by the pendulum method described by HÖLZL and SCHRAMMEN /3.35/.

Common to all these methods is the fact that the current arising from the change in capacitance is an alternating current which can be easily detected. When used with a phase sensitive detector (PSD) automatic adjustment of the compensation voltage can be achieved. The accuracy obtainable with such arrangements, which is

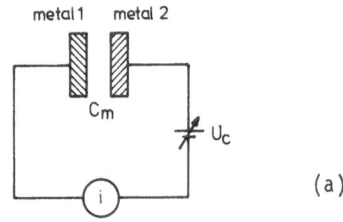

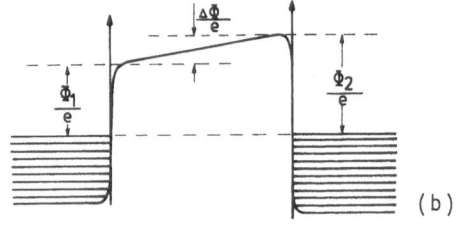

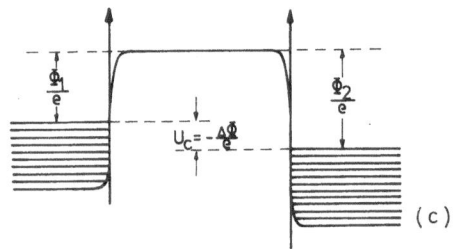

Fig. 3.13.   a) Basic principles of the vibrating capacitor method:
              C_m: experimental capacitor with variable capacitance
              U_C: variable voltage source for bucking voltage
              i: ac current detector

              b) Potential variation between the condenser plates when $U_C = 0$

              c) Potential variation between the condenser plates when $U_C = - \Delta\Phi/e$

dependent on the care with which the electronic, and in particular the mechanical,
parts of the circuit are constructed, can be ±0.1.....±10 meV. The condenser must
be field-free during the course of the measurements.

   If the geometrical size of the apparatus (electrode size and spacing) is large
in comparison with the patches for polycrystalline samples, then for such samples
the CPD is given by

$$\Delta\Phi/e = (\bar{\Phi}_1 - \bar{\Phi}_2)/e. \tag{3.13}$$

Here $\bar{\Phi}_1$, and $\bar{\Phi}_2$ are the average work functions of the electrodes, according to the relation $\bar{\Phi} = \sum_i f_i \Phi_i$; f_i represents the fraction of the target area having a WF Φ_i.

b) Systematic Sources of Error

The reliability and reproducibility of the results are basically limited by the lack in constancy of the work function of the reference electrode during the measurements (for example, because of gas adsorption), and by parasitic electrical fields between the electrodes of the condenser and the surrounding parts of the apparatus (stray capacitance).

Frequently the work function changes of the target are determined by adsorption processes. If the adsorbate is a metal, which is evaporated from a localized source, then the reference electrode can be screened by suitable methods from bombarding atoms. Under ultrahigh vacuum (UHV) conditions the work function in this case can be relied on to be constant. If, however, gas adsorption processes are to be investigated, then such screening of the reference electrode is in general not possible. In this case the material of the reference electrode must be chosen so that its work function is inert with respect to the adsorbate. The uncertainty caused by the reference electrode in such investigations should on no account be underestimated.

The influence of stray capacitance on CPD measurements has been thoroughly investigated in the past /3.36,37/. Through the condenser electrodes not only is the capacitance of the "Kelvin condenser" varied, but also the capacitance between the vibrating electrode and various other surrounding parts of the apparatus. As each of these parts has a different work function, so additional signals are created which cannot be distinguished from the true Kelvin signal. Systematic investigations of this problem have shown that the measured CPD can, in unfavorable cases, vary by several hundred mV for changes in the vibration amplitude or the equilibrium separation of the condenser plates /3.37/.

In order to obtain reliable results, it seems to be necessary to hold both equilibrium spacing and vibration amplitude constant and reproducible, and to connect the fixed, not the vibrating electrode, to the input of the "detection system", as well as to use the vibrating electrode as the "reference". This has two basic advantages; first, the modulation of the stray capacitance and hence the perturbing spurious signal is much smaller at the fixed electrode than at the vibrating electrode; second, the WF changes caused by means of adsorption processes on the target (fixed electrode) are compensated by the bucking voltage, so that the stray fields remain unchanged during the experiment. By taking this precaution, although the absolute value of the CPD measured between target and reference electrode may still be erroneous, at least it is not of the same magnitude as the change in the CPD!

c) Detection System

Originally the Kelvin signal was amplified and the compensation voltage U_C adjusted
manually for signal minimum to determine the CPD. U_C and hence $\Delta\Phi$ could then be read
from a voltmeter /3.27/.

Today most systems use a phase sensitive detector (PSD), which feeds back the
rectified voltage to the condenser and so automatically sets the compensating voltage to the signal minimum. Thus, the CPD can be continuously followed and recorded
on an X - Y plotter.

Such systems have been described by SIMON /3.38/, DELCHAR, and EHRLICH /3.39/,
and PETIT-CLERC and CARETTE /3.40/.

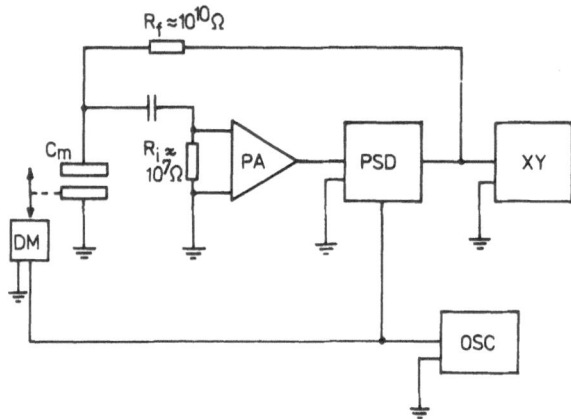

Fig. 3.14. Self balancing circuit for the CPD measurement: C_m: experimental capacitor, excited to vibration by the oscillator OSC and the driving mechanism DM. The
resultant ac current gives rise to a voltage drop across R_i, the input resistor of
the preamplifier PA. The dc output of the phase sensitive detector PSD is fed back
to C_m via R_f as well as plotted by the recorder XY

The schematic circuit diagram of such a system is shown in Fig. 3.14. The individual
components are listed in the figure caption. BLOTT and LEE /3.41/ have described a
system which is basically different from all others in that one electrode vibrates
at two frequencies simultaneously. Hence, at the input of the detection system, one
frequency component is amplitude-modulated by the other. With this particular system
it is possible to make CPD measurements between the electrodes even when the fixed
electrode emits electrons or ions, or when the vibrating electrode is connected to
the input of the measuring system.

We shall not attempt here to go into the electronic details of detection systems
although one general comment should be made: all the preamplifiers mentioned in the

literatur for the Kelvin signal employ a high input resistance R_i, in order to achieve a high voltage drop by passage of the a.c. current from the Kelvin condenser. In order that the signal does not flow through the feedback resistance R_f, this must necessarily be greater than R_i. Typical values of R_i and R_f are $R_i \approx 10^6 \ldots 10^7$ Ω and $R_f \approx 10^8 \ldots 10^{10}$ Ω . A few problems can arise because of these high resistances. The capacitance of the leads to the electrodes must be as small as possible so that the signal is not reduced, and the insulation of the high impedance electrode must be extremely good. Electron or ion currents which enter or leave the high impedance electrode give rise to a voltage drop across R_f which causes erroneous measurements. In the particular case of CPD measurements with a hot target which are being carried out in our laboratory at present, there are considerable problems with both the insulation and the thermal emission of the target. Good results can be obtained by means of a preamplifier with a virtually grounded input. The current from the Kelvin condenser gives rise to a voltage drop across the feedback resistance of the operational amplifier, Fig. 3.15.

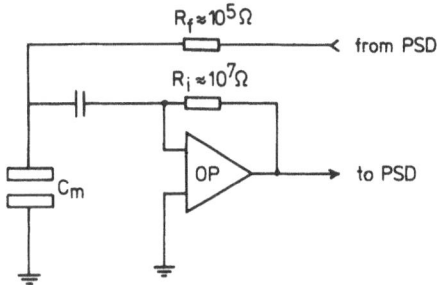

Fig. 3.15. Input stage of the self-balancing circuit, using the virtually grounded input of an operational amplifier OP as proposed by the author. (See text)

This voltage drop, which is just as large as in conventional circuits since $R_i = 10^7$ Ω , can be used for further processing. As a consequence of the low impedance the feedback resistance for the compensation voltage needs to have a value of only about 10^5 Ω ; thus parasitic resistance in the target insulation of the order $\gtrsim 10^6$ Ω and parasitic currents of the order $\gtrsim 10^{-8}$ A do not disturb the measurements.

d) <u>Form of the Kelvin Method in Practice</u>

Following are brief descriptions of 3 experimental arrangements, of which the first is a modern version of the classical Kelvin method, and the second and third have been designed specially to satisfy particular experimental requirements.

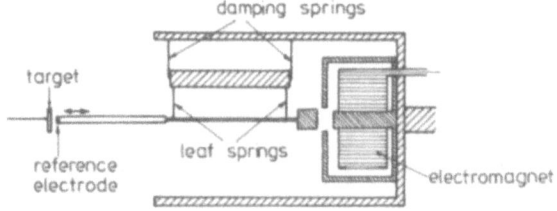

Fig. 3.16. "Classical" Kelvin method as described by JASCHINSKI /3.42/

JASCHINSKI /3.42/ has described a "classical" arrangement which was used to measure the work function of a molybdenum target during Ag deposition, Fig. 3.16. The target, Mo (100), is very close to the front face of a molybdenum rod of 2 mm diameter which serves as the reference electrode, and which vibrates along its long axis. The vibration excitation is effected by means of UHV electro-magnet, the mechanical resonance frequency of the system being determined by means of leaf springs. Transfer of vibrations to other parts of the system is largely prevented by damping springs (prevention of microphonic pick-up). The system works at a resonance frequency of 380 Hz, an equilibrium spacing of 0.5 mm between target and reference electrodes and vibration amplitude of 0.1 mm. The zero compensation of the compensating voltage is performed with the aid of a phase sensitive detector and can be carried out to an accuracy of ±1 mV.

In such a classical method bombardment of the target with foreign atoms cannot take place simultaneously with CPD measurement. The target must therefore be brought in front of the evaporation oven, and back again for CPD measurement, with the aid of a manipulator. Hence the dependency of the work function of the target on the adsorbate coverage can only be performed stepwise. The reproducibility of the measurement is considerably limited by this manipulation (e.g., the necessary constancy of the inter-electrode spacing could not be maintained). JASCHINSKI has given the uncertainty arising from this procedure as ±10 meV.

HÖLZL and SCHRAMMEN /3.35/ on the other hand recently have described a "pendulum device" arrangement which enables a continuous observation of the work function change of a target while it is bombarded with a beam of atoms. In this method the variation of capacity between target and reference electrode is brought about by a periodic oscillation of the reference electrode in a plane parallel to the target surface. The basic principle of this device, as shown in Fig. 3.17, consists of two main elements: first, the target, surrounded by a guard ring for minimizing effects of stray fields; second, the pendulum, made of some elastic material (leaf spring, elastic rod, etc.) and the reference electrode. The oscillation of the pendulum is provided by a driving mechanism (e.g., electromagnet) at the mechanical resonance frequency of the system.

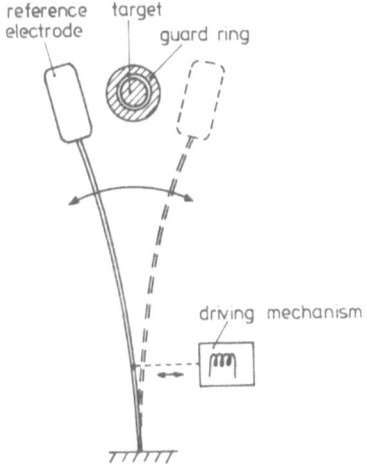

Fig. 3.17. Principle of the "pendulum device". /3.35/

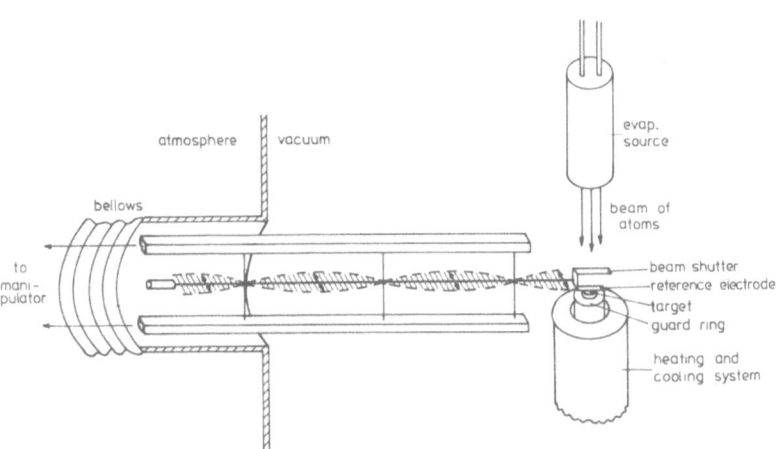

Fig. 3.18. Sketch of a more recent "pendulum device" system in order to study simultaneously the WF variation in adsorbate systems. /3.43/

Fig. 3.18 shows an arrangement as it is used at present in our laboratory in
order to study the adsystem Ni/Ni at various temperatures /3.43/.

In this experimental arrangement the pendulum consists of a long molybdenum rod
(1 mm in diameter x 400 mm length). This rod is supported in three positions by
thin torsion wires and is induced to perform oscillations in the second harmonic
mode (shown as dotted lines; resonance frequency, 64 Hz) with mechanically fixed
points corresponding to the nodes. With the use of such a slightly varied "pendulum
system the transfer of vibrations to other parts of the apparatus is minimized. In
addition, since the reference electrode is mounted at one end of the molybdenum rod
while the lump of iron necessary to produce the mechanical excitation of the system
is mounted at the other, effective shielding with respect to the magnetic field is
possible.

The target (together with its guard ring) is mounted about 0.1 mm below the plane
of oscillation of the reference electrode. (In the figure this spacing is exaggerated
for the purpose of clarity). As the diameter of the target and the oscillatory of
the reference electrode are 1.5 mm and 10 mm respectively, the beam of atoms impinging on the surface of the specimen can, with respect to time, be regarded as
nearly continuous.

The reference electrode itself must be screened against the atomic flux by means
of a beam shutter fixed to the oscillating rod. Suitable electronic facilities permit
a measurement of the work function change of the target with an accuracy of ±1 meV
to be recorded during continuous deposition on it.

BUTZ and WAGNER /3.29/ have described a variation on the Kelvin method, by which
local WF variations on the surface of a sample could be observed with high spatial
resolution along one coordinate. They use their method to determine concentration
profiles in the case of surface diffusion of oxygen on tungsten (110).

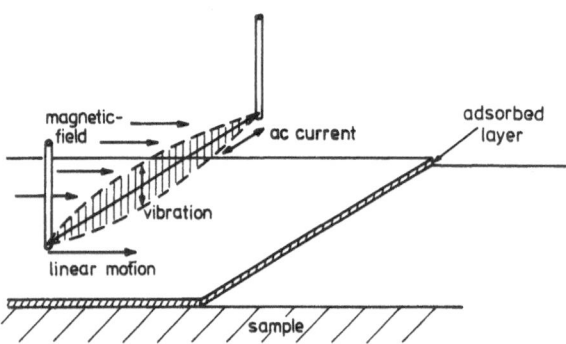

Fig. 3.19. Schematic drawing of the device used by BUTZ and WAGNER /3.29/ for CPD
measurements with high spatial resolution

A wire of 6 μm diameter serves as the vibrating reference electrode and is mounted parallel to the target surface at a spacing of about 15 μm, Fig. 3.19. The vibration of the wire perpendicular to the surface of the sample is effected by means of the interaction of an alternating current passed through the wire with the field of a permanent magnet. Wire and sample can be shifted parallel to each other, so that work function changes along the coordinate perpendicular to the wire can be registered. The spatial resolution of this arrangement is given by authors as about 50 μm, and the resolution of work function difference as ±20 meV.

e) Other Condenser Methods

A few other condenser methods are described in the literature which do not adhere to the vibration principle of Zisman. In fact they clearly have little application, but two recent examples are mentioned here.

FAIN et al /3.44/ have described a device which consists of a flat sample and a ribbon parallel to its surface (reference electrode distance r). A bucking voltage is applied between these two electrodes which is made up of a dc voltage (U_b) and an alternating voltage ($U_o \sin\omega t$). The resulting electrostatic attraction force F_r causes the ribbon to vibrate according to the relation

$$F_r = (1/2) \ (U_b + \Phi/e + U_o \sin\omega t)^2 dC/dr. \tag{3.14}$$

If U_b is adjusted so that $U_b = \Delta\Phi/e$, then the component of frequency ω in the Fourier analysis of the force F_r, and hence also in the vibration of the ribbon, disappears. The disappearance of this frequency component is used for the adjustment of the bucking voltage. This method is superficially very similar to the Kelvin method, but in principle it differs considerably. This can be seen easily since in the case of the Kelvin method the displacement current is used to measure the CPD, while in the above method the imbalance of the capacitance bridge produced by the oscillation of the electrode is used. Note that one branch of the capacitance bridge is made up of the experimental condenser.

A completely different arrangement, which can however be included in the capacitance methods, is the one described by KRIMMEL et al /3.45/. The arrangement here comprises a linear electron source and an electrostatic biprism, consisting of two parallel grounded plates and a straight wire mounted between them, (see Fig. 3.20). When a small positive voltage U_F is applied to the wire, the electrons are split in two beams and give rise to interference fringes on the fluorescent screen. If the CPD between the wire and the plate changes, for example by coating the wire with a foreign layer, then the interference pattern changes. This change (a',a' → b',b') can be compensated by adjusting the voltage U_F.

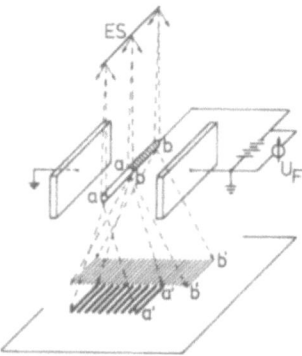

Fig. 3.20. Electron interferometer for measuring changes of CPD. ES: linear electron source: a-b: biprism filament, placed between two parallel grounded plates. (ES, U_F and a'a'; b'b' are explained in the text) /3.45/

In the figure, the relationships are depicted which arise when the two halves of the wire a-a and b-b have different work functions.

4. Work Function of Pure Metals with Clean Surfaces

J. Hölzl

In this chapter, where experimental results from pure metals with clean surfaces are reviewed, the following topics are discussed:

First, for purposes of continuity a brief (mainly qualitative) summary of various theoretical models is presented in Sect. 4.1, which is divided in two parts. The first part deals with empirical and semiempirical studies, principally the STEINER--GYFTOPOULOS model /4.14/ and the second with the more quantum-mechanical treatments; however, the latter is to be regarded only as a brief summary of the subject. Any reader interested in a comprehensive presentation of the theory (and in various theoretical details) is referred to the Sect. 2.2.3-6 in Chap. 2.

Second, in order to compare WF data calculated mostly from idealized models with those obtained experimentally with a real crystal, two major questions must be answered, namely:

(a) Is there any dependence of the WF data on the method of measurement used? Clearly the presence of patch effects, and of stray and/or external electrostatic or magnetic fields, etc., must be important here. Furthermore, the differences between the measured parameters, e.g., between the true WF and the so-called Richardson WF, must be appreciated. These and similar questions have been treated from the experimental viewpoint in Sect. 3.1,2.

(b) What kind of preparational procedure are necessary to optimize, in the case of a monocrystal sample, the crystallographic and geometrical ordering, and to mini-mize contamination of the sample surface? This will be discussed in Sect. 4.2.

In Sect. 4.3,4 variations of the WF as a consequence of external influence will be discussed. Thus in Sect. 4.3 various parameters governing the variation of WF with temperature are considered, and experimental results connected with first- and second-order transition processes are reviewed. (See Table 4.1). In Sect. 4.4 the mechanical stress dependence of WF is briefly outlined and a few experimental results are given in Table 4.2. Finally in Sect. 4.5 WF data are presented which cover nearly all metals. These data are compiled in Table 4.3.

4.1 Summary of Theoretical Models Used for the Caluclation of the Work Function of Pure Metals with Clean Surfaces

4.1.1 Empirical and Semiempirical Studies

There are many empirical and semiempirical approaches to the WF that correlate it with atomic, bulk and surface properties such as atomic number /4.1,2/, atomic ionization energy /4.3/, atomic volume /4.4/, crystalline packing density in con-nection with atomic ionization energy /4.5/, sublimation entropy /4.6/, surface energy /4.7/, and other properties /4.8-12/. As an example of these studies we wish to discuss the correlation with electronegativities /4.13-15/ in some detail, since it provides insight into the systematic trends of the WF's of the elements.

PAULING /4.16/ characterizes the electronegativity as "the power of an atom in a molecule to attract electrons to itself". In quantitative terms the electronega-tivity χ of a neutral atom is defined as the arithmetic mean of its first ionization energy, I, and its electron affinity, A /4.16a/

$$\chi = \frac{1}{2} (I + A). \tag{4.1}$$

Clearly the outermost layer of a metal which forms the surface plays a distinctive role. Hence, it may be expected to be a good approximation that this layer forms a special system of electronegativity χ with respect to the bulk material to which it is connected. GORDY and THOMAS /4.13/ established a relation between WF and electronegativity for the first time. They fitted WF data to electronegativity values with a straight line and found

$$\Phi = 0,72 \chi + 0,34, \tag{4.2}$$

where χ as well as Φ are given in eV. This relation describes roughly the trend

of the WF for polycrystalline surfaces of the elements, but there are also considerable deviations from it as can be seen from Fig. 4.1. According to GORDY /4.17/ and

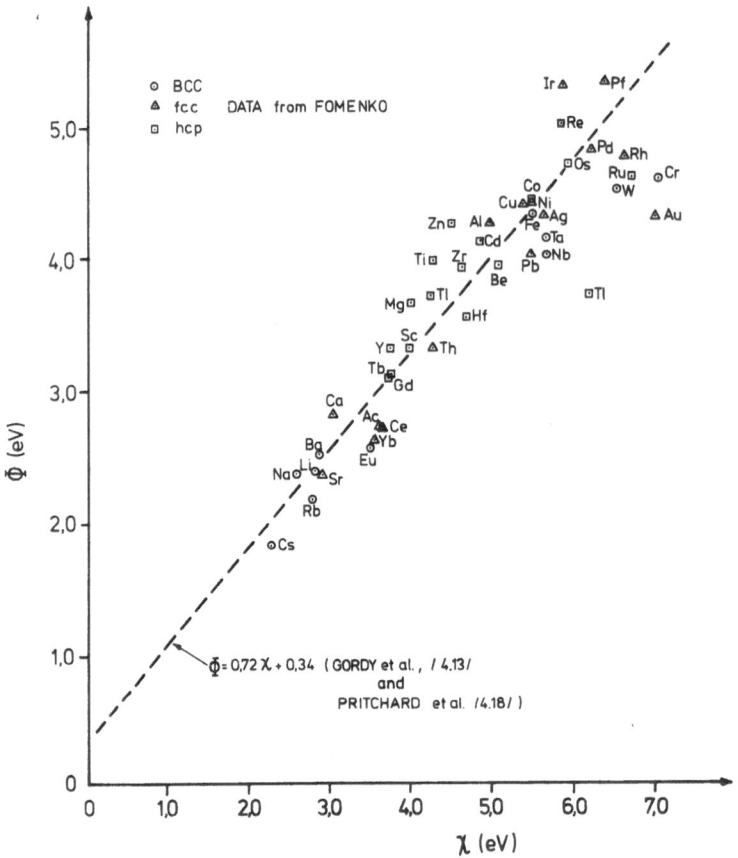

Fig. 4.1. Correlation between WF and electronegativity χ. /4.45,13,60a,18/

PRITCHARD and SKINNER /4.18/ the electronegativity χ of an atom in a molecule can
be expressed in terms of the number v of electrons per atom which participate in
bonding, and the effective radius r of the atom in the bonded state, so that

$$\chi = 0.98 \, \frac{v+1}{r} + 1.57, \tag{4.3}$$

where χ results in eV if r is expressed in Å. Fig. 4.2, taken from GORDY's paper
/4.17/, shows that this relation is obeyed surprisingly well by a large number of
elements with the exception of Au, Cu and Ag.

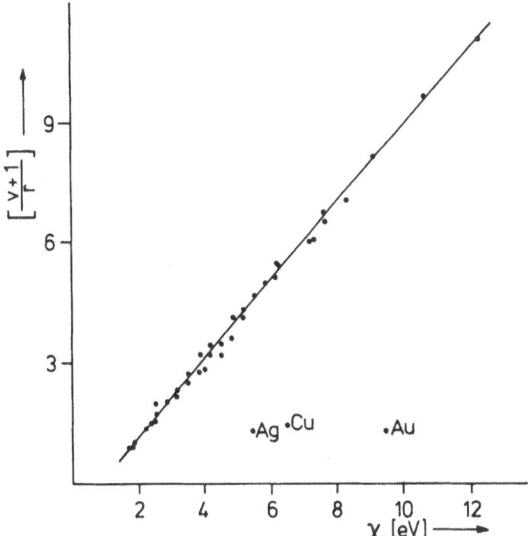

Fig. 4.2. Plot of electronegativity χ as a function of $(v + 1)/r$ for a great number of elements given in [Ref. 4.17, Table 1] the meaning of v and r is given in the text. /4.17/

Expression (4.3) has been extended to surface atoms by STEINER and GYFTOPOULOS /4.14/ who replace v by the quantity v_s, the number of electrons per surface atom which participate in bonding and analogously r by r_m, the radius of these atoms. Moreover, they postulate that the eletronegativity modified in this way may be identical with the WF of the respective metal, hence

$$\Phi = 0,98 \, \frac{v_s + 1}{r_m} + 1,57. \tag{4.4}$$

The surface valency v_s is related to the numbers of next, N_s', and next-nearest neighbors, N_s'', of surface atoms and to the respective fractional bond numbers n' and n" via $v_s = N_s' n' + N_s'' n''$. The fractional bond numbers can be obtained from the corresponding relation for the metallic valency v_m, which contains the number of next, N_b', and next-nearest, N_b'', neighbors in the bulk together with an empirical relation due to PAULING /4.16/ which associates n' and n" with the interatomic separations R' and R" of next and next-nearest neighbors, respectively, that is

$$R'' - R' = 0,26 \, \ln \, (n'/n''). \tag{4.5}$$

In this way the surface valency is given by

$$v_s = \frac{N'_s + \exp\left(-(R'' - R')/0{,}26\right) N''_s}{N'_b + \exp\left(-(R'' - R')/0{,}26\right) N''_b} \, v_m \, .$$

(4.6)

In Table 2.1 some data calculated from (4.4) and (4.6) are compared with experimental results. The agreement with experiment is very good. A somewhat different model, which allows for the calculation of semiempirical WF data, is suggested by ALBRECHT /4.8/. This rather crude approach consists of considering a gas of quasi free electrons with a Fermi energy ε_F in the volume part of the crystal. The behavior at the surface is described by strongly interacting atoms. Thus the WF is given as a function of crystal and atomic parameters. The data, obtained on the basis of ALBRECHT's model are in good agreement with experimental results for a great number of metals.

On the basis of quite a different semiempirical approach FRITSCHE and NOFFKE /4.19/ have studied very recently the difference of the WF of metals for various surface planes. In contrast to SMOLUCHOWSKI's earlier more qualitative study, /4.20/ these authors have obtained quantitative results in excellent agreement with experiments. They used a linear superposition of atomic charges which were calculated self-consistently by means of a program designed by LIBERMANN et al. /4.21/.

The semiempirical treatments mentioned above cannot really offer fundamental insight into the subject. However, such approaches can certainly be used as a starting point in the explanation of new experimental observations, as it was recently done, e.g., by KRAHL-URBAN /4.22/ in discussing the variation of WF's in the case of stepped surfaces.

In order to complete this theoretical survey a presentation of both older and more recent quantum-mechanical models is necessary. As these are reviewed comprehensively in Chap. 2, the theoretical part of this article, here only a brief outline of these treatments is given.

4.1.2 Outline of Quantum-Mechanical Treatments

The basic ideas about the quantum-mechanical calculation of the WF were developed by WIGNER and BARDEN /4.23/ roughly 40 years ago. Considerable progress was made recently by the application of the density-functional formalism /4.24/ to a uniform-background or jellium model /4.25/, and to models which include the lattice potential from the outset, /4.26-29/. To calculate the WF in principle one has to start from a Schroedinger equation for a system of N mutually interacting conduction electrons which move in the external potential $v(r)$ caused by the ion cores. For practical computation this many-body problem has to be reduced to a one-body form. To accomplish this, WIGNER and BARDEEN employed the Hartree-Fock approximation,

whereas in practically all modern calculations the density-functional formalism
has been used. In any case, to obtain reasonable results, it is necessary to take
the individual Coulomb interactions between the electrons into account via exchange
and correlation potentials.

Since the WF Φ (see its definition in Sect. 2.2) can be considered to consist of
two terms $\Delta\phi$ and $\bar{\mu}$, the surface and bulk contributions respectively, that is,

$$\Phi = \Delta\phi - \bar{\mu}, \tag{4.7}$$

then in general calculation of these two terms has to be done separately. Note that
$\bar{\mu}$ can be obtained from a self-consistent band structure calculation of the bulk,
whereas the determination of $\Delta\phi$ requires a self-consistent calculation of the dipole
barrier at the surface.

Extensive studies of the bulk part of the WF have been performed, mainly by HODGES
/4.30/. With the help of measured WF's he has calculated $\Delta\phi$ from (4.7).

Since a wave-mechanical self-consistent treatment of the surface electron density
requires a huge amount of computer time, extensive use has been made of the concep-
tually simple uniform-background model /4.25/. For simple free electron metals this
model gives satisfactory results, with further improvement possible by taking into
account perturbationally a pseudopotential correction to simulate lattice effects.
In the case of simple metals the WF's calculated in this way agree with measured
WF's to within 5 - 10 %. Note, however, that the applicability of this model is
restricted to nearly free electron metals; for d-band metals, for example, it is not
suitable. Very recently calculations have been carried out for the Li (100) /4.26/,
Na (100) /4.27/, Al (111) /4.28/, Nb (100) /4.29/, and Cu (100) /4.31/ faces, employing the lattice potential from the outset. For details the reader is referred
to the excellent review article by LANG /4.25/, or to the theoretical part of this
review.

In closing this short theoretical survey the reader is referred to Table 2.1 in
the chapter on Theory, where a comparison is made between measured WF's and results
obtained by the theories of STEINER-GYFTOPOULOS /4.14/, and of LANG-KOHN /4.32/,
and by different calculations which include the lattice potential from the outset.

4.2 Preparational Procedures

As it is well established that the WF is extremely sensitive to surface condition,
special care must be taken with respect to
    a) cleanliness of the surface, i.e., absence of contaminants and impurities;
    b) knowledge about the structure and perfection (smoothness) of the uppermost
       layers of the crystal.

Taking cleanliness first in order to produce a surface which is free from contaminants, the sample has to be subjected to one or more of a variety of treatments.
Of these, heat treatment, ion bombardment, and gas reaction followed by heating, are
the most commonly used. In may cases satisfactory results can be obtained using
simple treatments, as can be confirmed by AES or X-ray Photoelectron Spectroscopy
(XPS), but occasionally difficulties arise that need more complicated procedures.
In some cases, for instance, the macroscopic sample contains small amounts of contaminants in the bulk, which result from the inclusion of gases, or of carbon, sulphur, etc. It is very easy to eliminate the gaseous constituents by means of heat
treatment, but with carbon, sulphur, etc. considerable difficulties are encountered
when the diffusion of such constituents towards the surface is considered. To demonstrate this difficulty, the processes that have been used for cleaning the Ni
(111) surface may be of some interest. Two methods are well known: argon-ion bombardment and oxidation/reduction cycles (in O_2, followed by H_2, atmospheres). The
first is sometimes disadvantageous since during the ion bombardment an unexpected
carbon contamination can appear /4.33/, and moreover dramatic roughening of the
surface takes place. The oxidation/reduction procedure, however, has been shown to
be quite successful. In this treatment, as demonstrated by HEIMANN and HÖLZL /4.34/,
one of the main experimental precautions to take is to mount the sample, a highly
pure (99,9999), very thin, (d=0.5 mm) Ni (111) crystal plate, on the top of hair-
pin shaped rods which were themselves of highly pure polycrystalline Ni. In the
course of the heat treatment (T≈800°C), and the subsequent oxygen ($P_0 \approx 10^{-5}$ Torr
at T=700°C) and hydrogen ($P_{H_2} \approx 10^{-5}$ Torr at T=200°C) cycles, both the bulk carbon
and sulphur constituents were extracted so that after a great number of cycles no
surface contamination could be detected. To check the absence of both sulphur and
carbon on the surface AES was used. In the course of the experiment, however, it
was found that the ratio of the intensities of the surface plasmon loss (=9,0 eV)
to that of the volume plasmon loss (19,0 eV) still exhibited drastic variations
even when no AES contaminant signals could be observed. It is possible that such
plasmon loss variations might represent an even more sensitive tool for the monitoring of surface contamination than AES.

In the case of refractory metals such as W, Mo, etc., the situation seems to be
somewhat easier. As was demonstrated by BESOCKE /4.35/, heating a W monocrystal
for several hours at T=2000°C under UHV conditions was sufficient to remove bulk
gaseous inclusions. To achieve complete decarbonisation as well, it was necessary
to keep the crystal at the same temperature of 2000°C in 10^{-7} Torr of oxygen for
several hours. In the experiment only AES was used to monitor the presence of contamination at the surface.

With regard to structural considerations, it is clear that WF values, in addition
to their great sensitivity to small amounts of surface contamination, are also highly
sensitive to crystal structure and perfection. Thus the surface term $\Delta\phi$ in (4.7)

must be discussed first of all with respect to its dependence on crystal orientation; this can be understood in terms of the model proposed by SMOLUCHOWSKI /4.20/. In his approach two effects with respect to the electron density near the surface are considered. First, it is assumed that the positive charge within a Wigner-Seitz cell is smeared out continuously. As a consequence of minimizing the total energy of the system there exists a so-called spilling out of electrons perpendicular to the surface which results in a shift of the center of negative charge in an outward direction. This "spreading effect", which is practically independent of the packing density at the surface, causes a lowering of $\Delta\phi$. On the other hand there exists also a relaxation of the electron density parallel to the corrugated surface. This "smoothing effect" which results in a displacement of the center of electron charge density back towards the surface is strongly dependent on the packing density in the surface region and therefore on surface orientation. It is noticeable that the latter contribution is large for "open" and small for "closed" crystallographic structures. On applying SMOLUCHOWSKI's model the WF is found to decrease in the sequence (110) → (110) → (111) for the bcc lattice and to increase in the same sequence of these faces for the fcc lattice.

This theoretical result is in most cases in excellent agreement with experimental findings. A quantitative discussion of this topic is given in the chapter on Theory.

So far we have considered an idealized·crystal; when considering a real mono-crystalline sample the surface certainly cannot be assumed to be atomically flat and strictly periodic, as there exist surface defects of various kinds. HENZLER /4.36/ has recently reviewed the detection and evaluation of this geometrical per-turbation of the surface (zero-, one-, two-dimensional defects) by means of electron diffraction studies. The influence of these defects on WF have been in the immediate past a subject of intense investigation. PLUMMER and RHODIN /4.37/ have shown that a few tungsten atoms deposited onto a W (110) field emitter tip cause a WF variation according to $\Delta\phi/\phi_0 = 3,6\ N_a$ where N_a is the surface density of adatoms per $\overset{\circ}{A}{}^2$ and ϕ_0 is the WF of the "clean" surface. With regard to macroscopic crystals the effects of various defects have been studied. In the case of conventional cleaning, as per-formed by ion bombardment and subsequent annealing of the sample, WF reductions of several hundred meV can be observed. Effects of this kind were recently reported on aluminium /4.38/ and platinum /4.39/. A detailed study of this basic behavior has been made by KRAHL-URBAN et al. /4.40/. These authors studied the WF of regularly stepped W single crystal planes as a function of terrace width and step orientation. One of the main results of this experiment was that the WF decreased linearly with step density for a given step orientation. In Fig. 4.3 a plot of the true WF at 2300 K of the stepped surfaces (type: W(S)- [m(110)x(1$\bar{1}$0)])[1] versus the correspon-ding step density N_s is shown.

---

[1] The nomenclature of the stepped tungsten single crystal surface is given in the original paper /4.25/.

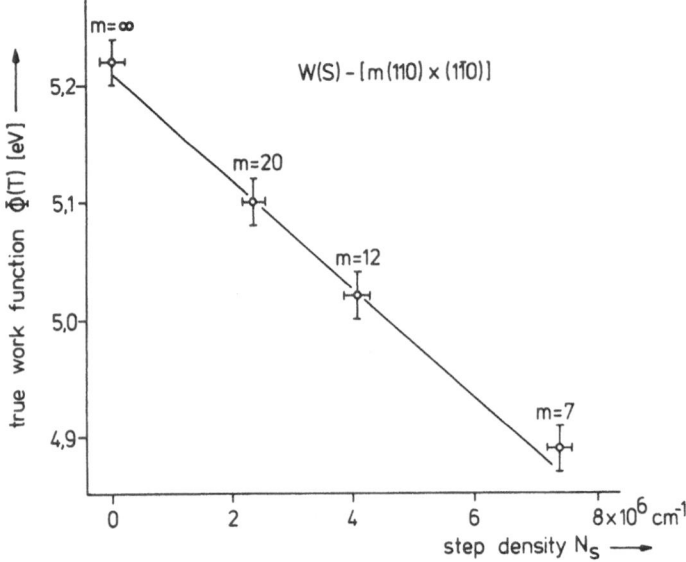

<u>Fig. 4.3.</u> True WF for T = 2.300 K of stepped surfaces type: W(S) - [m(110) x (1$\bar{1}$0)] versus the corresponding step density N_S. /4.40/

This finding is consistent with the SMOLUCHOWSKI smoothing effect of the electron charge distribution caused by the structural arrangement of surface atoms. Note that a review paper covering this topic is given in this volume /4.41/.

Quite a different way to check the surface smoothness of an Al monocrystal has been reported by GREPSTAD et al. /4.38/. There, the correlation between surface roughness and the photo yield due to surface plasmon decay /4.42/ was used. As a consequence of the dominance of the surface photo electric effect at low photon energies /4.43/ on an ideally smooth surface, the photo yield with p-polarized light Y_p at large angles of incidence will be several times larger than the yield with s-polarized light Y_S. On a real surface, however, the yield due to the roughness-assisted surface plasmon excitation and decay will tend to reduce the vector ratio Y_p/Y_S which, therefore, may also serve as an indicator of the smoothness of otherwise clean surfaces.

Finally, in connection with this highly important question it should be noted that an assessment based on numerous experimental results has been given by RHEAD /4.44/. He has suggested that the changes, when extrapolated to complete structural disordering, amount to about 10 percent of the perfect structure WF, and that a 1 percent defect concentration can produce a reduction, depending on the metal, of

the order of 10 - 20 meV. As can be seen in Chap. 3 this is well within the range
of detectability.

Since both cleanliness and the disappearence of possible crystal defects are
often closely connected with heat treatment of the sample, it is necessary now to
discuss a few temperature effects on the WF, in the next section.

4.3 Temperature Effects on the Work Function

As can be inferred from the definition of WF (see Sect. 3.1) a knowledge of $d\Phi/dT$
is of primary fundamental interest since it permits comparisons to be made between
CPD WF measurements and those involving thermionic emission. Since in the latter
experiments only the apparent WF $\Phi*$ can be obtained, by means of Schottky plots,
the calculation of the true WF Φ requires the knowledge of $d\Phi/dT$. Results for the
(110) face of W as measured thermionically by KRAHL-URBAN /4.22/ are
$\Phi* = (5.75 \pm 0.02)$eV, the apparent emission constant
$A* = (1733 \pm 180)$A cm^{-2}K^{-2},
$\Phi = (5.22 \pm 0.02)$ eV at T = 2300 K and
$d\Phi/dT = -(2.30 \pm 0.09) \cdot 10^{-4}$ eV/K.

According to (4.7) we obtain

$$\frac{d\Phi}{dT} = \frac{d}{dT} \Delta\phi - \frac{d}{dT} \bar{\mu}. \tag{4.8}$$

The temperature coefficient $d\Phi/dT$ can also be expressed by the thermal expansion
coefficient $\alpha_{th} = \frac{1}{V} (\frac{\partial V}{\partial T})_p$, since the experiments are performed at constant pressure.

$$\frac{d\Phi}{dT} = \alpha_{th} V (\frac{\partial \Phi}{\partial V})_T + (\frac{\partial \Phi}{\partial T})_V . \tag{4.9}$$

There are many physical effects contributing to the terms in (4.8) and (4.9), which
are extremely difficult to estimate. For a detailed discussion the reader is referred
to the review article by HERRING and NICHOLS /4.47/. Using more recent results for
the uniform-background model, which have been derived by LANG and KOHN /4.32/,
SCHULTE /4.46/ estimated the first term on the right of (4.9). For a series of
simple metals he found this contribution to vary between 0.0 k_B for Al and -1.7 k_B
for Cs, where $k_B = 0.862 \cdot 10^{-4}$ eV K^{-1} denotes the Boltzmann constant. This is to be
contrasted with the result of HERRING and NICHOLS, who predicted this contribution
to be positive. Further, the effect of lattice vibrations on the chemical potential
$\bar{\mu}$ is probably a negative contribution of the order of a few k_B, according to HERRING
and NICHOLS. The remaining effects should contribute only a fraction of k_B. In
summary, the estimates of the temperature coefficient of the WF give the order of

magnitude of the observations, but it is not yet possible to predict its sign for a specific metal.

Experimentally such measurements are quite difficult, since very small amounts of surface contamination (as well as preceding heat treatments) of the sample can alter drastically not only the magnitude of dΦ/dT but also its sign. Experiments of this kind thus have to be performed in a very carefully controlled manner.

GARTLAND et al. /4.48/ who investigated the WF of a Cu monocrystal sample as a function of temperature, observed that the dependence in the case of the (111) face was strongly dependent on the cleaning process as can be seen in Fig. 4.4. Curve (a)

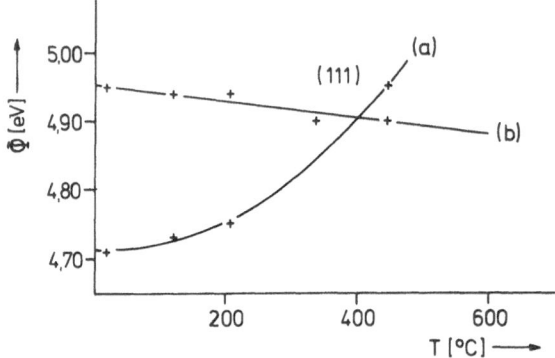

Fig. 4.4. WF vs. temperature measured on Cu (111)
curve (a): temperature dependence after 7 h of heat treatment
curve (b): measurement on a carefully cleaned surface. /4.48/

shows the temperature dependence after 7 h of heat treatment, while curve (b) refers to the measurement on the carefully cleaned surface. Measurements were performed during cooling from 800°C, and the estimated uncertainties were $\delta\Phi = \pm 0.03$ eV and $\delta T = \pm 50°C$. Their result was $(d\Phi/dT)_{exp} = 6 \cdot 10^{-5}$ eV/degree for the Cu (111) plane, in agreement with KÜHLER's experiment /4.49/.

Some other results of the change of WF with temperature are also of considerable interest. These are connected closely with certain mainly bulk transition processes, for example, the classes

(a) liquid/solid transitions
(b) crystallographic phase transitions
(c) super/normal conductivity transitions
(d) ferromagnetic/paramagnetic transitions.

To class (a): BUSCH et al. /4.50/ have studied the photoemission of both solid and liquid mercury. They found in that experiment that any possible discontinuity in the WF vicinity of the melting point must be smaller than 0,01 eV. This result is in agreement with a somewhat similar observation obtained on solid solutions of alkali metals which have been investigated by MALOV et al. /4.51/, and which will be mentioned in Chap. 6.

To class (b): The change of $\Phi(T)$ as a consequence of a bulk structural transformation has been studied by HILL et al. /4.52/ in the case of the $\alpha \rightleftharpoons \gamma$ transition of iron. Since α Fe and γ Fe have bcc and fcc structures, respectively, the corresponding WF's Φ_α and Φ_γ should be different, and there should be a change in WF when passing through the structural transformation temperature $T_{\alpha/\gamma}$ (=1183 K). They obtained from a thermionic experiment a difference $\Phi_\gamma - \Phi_\alpha$ of about 0.09 eV, while a positive ion emission experiment gave the same difference as about 0.06 eV.

To class (c): In a very recent experiment SCHOTT and WALTON /4.53/ have measured the contact potential difference (CPD) between tin and copper in the vicinity of superconducting transition temperature (T_{sc}=3.73 K) using the Kelvin method. They observed (in most of the experimental runs) no evidence of any sharp discontinuity in CPD at T_{sc} to within $\pm$ 0.2 mV. On the other hand all experiments showed a change in the value of d(CPD)/dT on passing through T_{sc}. Taking into account a noise level of only $\pm$ 0.2 mV, the slope of a well fitted straight line (averaged over many experiments) gave the result (2.2 $\pm$ 0.2) mV degree^{-1} over the interval 3.5 K to 3.73 K, while above T_{sc}, up to 4.2 K, the CPD was found to be essentially temperature independent.

To class (d): In an earlier paper CARDWELL /4.54/, using a polycrystalline Ni target, explained the variation of photocurrent vs. temperature by means of an anomaly in the WF at the Curie point T_c. Sometime later COMSA et al. /4.55/ measured directly the temperature dependence of Ni (polycr.) by the use of an electron beam method (CPD). They found a slight difference in the linear dependence of dΦ/dT above and below the ferromagnetic transition temperature:

$$\Delta\left(\frac{d\Phi}{dT}\right)_{\text{ferro para}} = (-0.99 \pm 0.17)\ 10^{-5}\ \text{eV/degree.}$$

However no jump in Φ as predicted later theoretically by PANT and RAJAGOPAL /4.56/ could be observed. COMSA et al. try to explain their results by means of a theoretical model given by WONSOWSKI and SOKOLOV /4.57/ which stated that

$$\Phi = W - \varepsilon_F \left(1 + \delta_1 M_{sp}^2\right), \tag{4.10}$$

where W, and ε_F are interpreted as the potential depth for the s-electrons, and their Fermi energy, respectively, δ_1 is a simple function of a few parameters incorporated in the above theory. M_{sp} denotes the spontaneous magnetization of the

bulk. In a more recent experiment by HÖLZL and PORSCH /4.58/ the study of COMSA
et al. has been repeated. In the latter investigation, using the pendulum device
method (see: Subsec. 3.3.2d) , a very pure surface could be obtained by coating the
Ni target with very thin layers of Ni films before starting each run. Numerous pre-
-experiments showed that spurious surface contamination, by e.g. sulphur, severely
affected the measurements. The result obtained was $d\phi/dT = -(0.15\pm0.01)$meV/degree
within the temperature range $230° \leq T \leq 450°C$.

Applying WONSOWSKI's theory HÖLZL et al. /4.58/ have given a possible explanation
in terms of the fact that not only the bulk but the surface magnetization of the
sample has to be considered. Nearly simultaneously CHRISTMANN et al. /4.59/ studied
the variation of WF with temperature of clean Ni(111) and (100) between 25 and 320°C,
and up to 430°C of Ni(110). In all cases the WF decreased linearly with temperature
by $1.7 \ 10^{-4}$ eV/degree. No anomaly (neither a jump nor a change of the slope of
$d\phi/dT$) was found at T_c, just as in the HÖLZL - PORSCH experiment. In agreement with
Ref. /4.58, 4.59/ photoemission experiments on Ni showed no WF change when passing
through T_c /4.114/. In contrast, the WF ϕ of metallic, i.e. heavily n-doped ferromagnetic semiconductors was found to vary drastically upon magnetic ordering. Differences $\phi_{ferro} - \phi_{para}$ of 0.1 - 0.5 eV have been observed with La- and Gd-doped
EuO /4.121/, and n-type $CdCr_2S_4$ /4.129/.

Finally in this Section, the temperature coefficients $d\phi/dT$ of a few metals in
some interesting temperature ranges have been collected in Table 4.1. Remarks about
the relevant transition processes have been included.

4.4 Mechanical Stress Dependence of Work Function

It has been established by several authors /4.61-63/ that there is a certain influence on the chemical potential $\bar{\mu}$ due to the elastic-deformation-induced change
in the volume of the crystal lattice. As a consequence of this, when using (4.7)
it follows at once that there must be a variation in WF. Assuming for the moment
that $\Delta\phi$ is constant (that is, there is no variation in the surface condition) then
$\Delta\Phi$ is expected to be positive or negative if $\bar{\mu}$ is decreased or increased, respectively,
by such a change.

Moreover there are additional contributions to the change in Φ, not only in the
elastic range but also above it, due to the increase in the concentration of imper-
fections in a crystal as discussed recently by MINTS et al. /4.64/.

During recent experiments by MINTS et al., who stretched various metals (Pb, Cu,
Au, Ag, Ni, Pt, Pd) unaxially at a rate of 2,5 mm/min, $\Delta\Phi$ was observed always to be
positive and the same for all metals ($\Delta\Phi = 1....5$ meV) in the elastic range of de-
formation. In the plastic range, however, the dependence of WF on strain ε_{st} always
showed a negative slope, with $\Delta\Phi$ ranging from 25.......200 meV. Some data from

Table 4.1

Metal	Temp.range (K)	$d\Phi/dT(eV \cdot K^{-1})$	Methods	Remarks	Reference
Hg	$234.29 < T < 273.16$	$+(4 \pm 0,7) \cdot 10^{-4}$	PE	liquid state (*1)	/4.60/
In	$429.77 < T < 618.16$	$-(2.4 \pm 0.5) \cdot 10^{-4}$	PE	liquid state (*1)	/4.60/
Ga	$302.94 < T < 623.16$	$-(1.20 \pm 0.25) \cdot 10^{-4}$	PE	liquid state (*1)	/4.60/
Sn	$3.5 < T < 3.75$	$+(2.2 \pm 0.2) \cdot 10^{-3}$	CPD	slight anomaly of $d\Phi/dT$	/4.53/
	$3.75 < T < 4.2$	≈ 0	CPD	at $T = T_{Sc}$ (*2)	
Ni(111) (100);	$298.16 < T < 593.16$	$-1.7 \cdot 10^{-4}$	CPD	no anomaly of $d\Phi/dT$ at $T=T_C$ (*3); Accuracy in	/4.59/
Ni(110)	$298.16 < T < 703.16$	$-1.7 \cdot 10^{-4}$	CPD	$\Delta\Phi = \pm 3 \cdot 10^{-3}$ eV	/4.59/
Ni(poly)	$503.16 < T < 723.16$	$-(0.15 \pm 0.01) \cdot 10^{-3}$	CPD	no anomaly of $d\Phi/dT$ at $T=T_C$	/4.58/
Cu(111)	$293 < T \lesssim 717$	$-(10 \pm 6) \cdot 10^{-5}$	PE	-	/4.48/
Cu(110)	$293 \lesssim T \lesssim 798$	$-(20 \pm 10) \cdot 10^{-5}$	PE	-	/4.48/

(*1) The WF of Hg, In and Ga varies in the transition liquid/solid (polycr.) by about 0.1 eV;

(*2) T_{Sc} means the superconducting transition temperature;

(*3) T_C means the ferromagnetic Curie temperature.

MINTS et al., unfortunately obtained under poor vacuum conditions ($5 \cdot 10^{-5}$ Torr), for bulk pure metals corresponding to ε_{st} = 25%, are listed as follows:

Metal (purity 99.999%)	$- \Delta\Phi$[meV]
Cu	28 ± 5
Au	60 ± 10
Ag	110 ± 30
Pt	100 ± 25
Pd	56 ± 10

Other experiments of this kind were performed earlier by BEAMS /4.65/, CRAIG /4.66/. CRAIG and RADEKA /4.67/, FRENCH and BEAMS /4.68/ and COHEN et al. /4.69/ under various conditions.

Most of the results obtained are in general agreement with a prediction given by DESSLER et al. /4.63/. In this approach the influence of a gravitationally induced electric field in conductors is studied which gives rise to a variation of the chemical potential and thus the WF of the conduction electrons due to the differential compression of a metal by gravity. Prior theoretical work on this field by SCHIFF and BARNHILL /4.70/ predict however a quite opposite result. A review paper on this topic has recently been published by J. STRNAD /4.73/.

Experimentally the situation is still unsatisfactory, in many investigations the surface conditions of the probes are poor or unknown, as commented by FRENCH and BEAMS /4.68/. LENERS et al./4.71/ have measured the compressionally induced change in CPD for the (100) face of a single crystal of copper. The cleaning of the surface was performed by Ar ion bombardment and measurements were made at a residual gas pressure of $2 \cdot 10^{-8}$ Torr. For a change in stress of about 34 atm these authors observed a variation of CPD vs. pressure of (-0,2 ± 1.2) µV/atm which was interpreted as a corresponding decrease in WF. Although this very interesting problem cannot be discussed fully in this article, a summary of the theoretical and experimental results of the gravitationally induced electric field outside of a vertical copper cylinder according to LENERS et al. /4.71/, is given in the following Table 4.2. There $\delta\Phi/e\delta y$ is listed, with references, and details of vacuum conditions, quality of cleaning and theoretical models; y is the vertical coordinate measured downward from the top of the cylinder.

Table 4.2

Experimental workers (year)	Studied plane	$\delta\Phi/e\delta y$ [μV/m]	Vacuum	Quality of cleaning	Reference
WITTEBORN and FAIRBANK (1967)	-	$-5.47 \cdot 10^{-5}$	10^{-11} Torr	No ultrahigh vacuum (UHV) cleaning; surface at 4.2 K so well covered	/4.72/
CRAIG (1969)	-	+4 to 6.4	Atmos.	Washed in alcohol	/4.66/
BEAMS (1968)	-	+1	$10^{-5} - 10^{-6}$ Torr	No vacuum cleaning	/4.65/
FRENCH and BEAMS (1970)	-	$+1.0 \pm 0.4$	Atmos.	Noted significant effects due to different cleaning procedures	/4.68/
BROWN et al. (1971)	-	-0.03 0.15	Atmos.	None reported	/4.74/
SCHUMACHER et al. (1972)	(110)	-7	$<10^{-10}$ Torr	UHV heating	/4.75/
LENERS et al. (1972)	(100)	-1.6 ± 1.0	$2 \cdot 10^{-8}$ Torr	Ar* bombardment (300°C high-vacuum bake)	/4.71/

Table 4.2 (Continued)

Theoretical workers (year)	Studied plane	$\delta\Phi/e\delta y$ [µV/m]	Theoretical model	Reference
SCHIFF and BARNHILL (1966)	–	$-5.6 \cdot 10^{-5}$	Shift due to gravity acting on a electron gas	/4.70/
DESSLER et al. (1968)	–	-0.6	Shift due to gravity acting on the crystal lattice	/4.63/
RIEGER (1970)	–	+0.17	Shift due to gravity acting on the lattice including electron-phonon interactions	/4.76/
LENERS et al. (1972)	(100) (110)	+3.6 ⎱ -3.2 ⎰	Semiempirical but includes surface-compression effects	/4.71/

4.5 Compilation of Work Function Data on Pure Metals

Table 4.3 gives a compilation of WF data covering nearly all metals. These data
were taken from recent work as far as possible. The results quoted have not primarily
been selected according to their relative merits but rather with the purpose of
providing the reader with unbiased guidance to the extensive literature. The Table
is arranged as follows:

Column 1,   Symbol of element; specified plane given in parentheses;

Column 2,   Work Function data in eV; errors are quoted as far as possible;

Column 3,   the method of measurement is indicated, viz
CPD - contact potential difference
PE - photoemission
TE - thermionic emission
FE - field emission;

Column 4,   year of publication;

Column 5,   reference.

5. Work Function Changes Induced by Adsorbates on Clean Metals

J. Hölzl

Many experimental studies of adsorption of both metallic and nonmetallic atoms on
various single (and polycrystalline) crystal substrates have been performed. These
studies can be divided into 'macroscopic' and 'microscopic' investigations. 'Macroscopic' methods deal with samples of an effective area of more than say 1 mm^2 whereas
'microscopic' studies are restricted to the emitting area of field emission tips.
A large number of different methods is already available for the study of surface
characteristics (e.g., coverage, dipole layers, etc.) of coated monocrystalline
surfaces, and the most effective method of studying any individual surface is to
combine several of these methods in situ. The WF and its change during gas or metal
exposure plays a more or less important role when analysing electron and ion spectroscopy data /5.1/. For experiments concerned with the measurement of WF change
most workers proceed as follows:
A carefully prepared substrate is coated with an adsorbate at a fixed temperature
T (isothermal adsorption) and the WF change $\Delta\Phi = \Phi - \Phi_s$ is monitored as a function

Table 4.3

Element	Work function [eV]	Technique	remarks	Year	Reference
Ag	4,0 ± 0,15	PE	polycrystalline film	1970	4.83
	4,26 ± 0,02	PE	–	1975	4.125
Ag(100)	4,64 ± 0,02	PE	–	1975	4.125
Ag(110)	4,52 ± 0,02	PE	–	1975	4.125
Ag(111)	4,74	PE	–	1973	4.126
Al(111)	4,26 ± 0,03	PE	fresh single crystal surfaces prepared	1973	4.82
(100)	4,20 ± 0,03		by autoepitaxy		
(110)	4,06 ± 0,03				
(poly)	4,28 ± 0,01				
Al(100)	4,41 ± 0,03	PE	–	1976	4.38
(111)	4,24 ± 0,02				
(110)	4,28 ± 0,02				
Au	5,1 ± 0,1	PE	polycrystalline film	1970	4.83
Au(100)	5,47	CPD	reference electrode tin oxide	1975	4.127
(110)	5,37				
(111)	5,31				
Ba	2,52	CPD	films evaporated on W(112) at T=77 K	1973	4.106
Be	4,1*	FE	thick film evaporated on:	1976	4.117
	5,3**		*:W(110); **:W(111) and ***:W(211)		
	6,0***				

Element	Work function [eV]	Technique remarks		Year	Reference
Be	4,98 ± 0,10	PE	–	1974	4.128
Bi	4,34 ± 0,05	PE	measurement done at 300°C	1974	4.105
Ca	2,87 ± 0,06	PE	constant φ value for film thickness ≳ 10 nm	1971	4.108
Ca	2,9	PE	film	1971	4.109
Ca	2,89	–	–	1971	4.107
Cd	4,08 ± 0,02	CPD	film on Tantalum substrate; reference: W.F. Barium	1955	4.118
Ce	2,9 ± 0,2	PE	polycrystalline film	1970	4.83
Co	5,0 ± 0,1	PE	polycrystalline film	1970	4.83
Cr	4,5 ± 0,15	PE	polycrystalline film	1970	4.83
Cs	1,95	PE	T = 25°C	1974	4.51
Cs	2,00*	CPD	films in the high coverage limit for substrates:	1971	4.111
	1,82**		$*$ /Ta(110)/; $**$ /W(100)/; reference values: Φ°_{Ta} (110) = 4,94 eV; Φ°_{W} (100) = 4,65 eV are assumed.		
Cs	1,88	PE		1971	4.107
Cu	4,65 ± 0,05	PE	polycrystalline film	1970	4.83
Cu(100)	5,10 ± 0,05	FE	– –	1973	4.91
Cu(110)	4,48 ± 0,03	PE	– –	1972	4.84
Cu(112)	4,53 ± 0,03				

Table 4.3 (continued)

Element	Work function [eV]	Technique	remarks	Year	Reference
Cu(100)	4,59 ± 0,03				
Cu(111)	4,94 ± 0,05				
Cu(110)	4,48 ± 0,03	PE	--	1973	4.48
Cu(112)	4,53 ± 0,03				
Cu(100)	4,59 ± 0,03				
Cu(111)	4,94 ± 0,03				
Er	$2,97 ± 0,65×10^{-4}$ T	TE	temp. range 1150-1500 K	1967	4.123
Eu	2,5 ± 0,3	PE	polycrystalline film	1970	4.83
Fe	4,5 ± 0,15	PE	polycrystalline film	1970	4.83
Fe(100)	4,67 ± 0,02	PE	-	1972	4.124
αFe(111)	4,81 ± 0,02	PE	-	1969	4.86
Ga	4,35 ± 0,05	PE	measurement done at 200°C	1974	4.105
Ga	4,30 ± 0,01	PE	at T = 50°C (liquid state)	1976	4.60
Ga	$4,32 ± 0,01 \leq \Phi$ $\leq 4,39 ± 0,01$	PE	at T = 0°C	1976	4.60
Gd	3,1 ± 0,15	PE	polycrystalline film	1970	4.83
Gd(poly)	2,90 ± 0,06	CPD	reference electrode: Au Φ_{Ag} = 4,30 eV is used.	1970	4.99

Element	Work function [eV]	Technique	remarks	Year	Reference
Hf	3,9 ± 0,1	PE	polycrystalline film	1970	4.83
Hg	4,475 ± 0,01	PE	T = -30°C (liquid state)	1976	4.60
Hg	4,43 ± 0,01 ≤ Φ ≤ 4,50 ± 0,01	PE	T = -100°C	1976	4.60
In	4,08 ± 0,04	PE	measurement done at 200°C	1974	4.105
In	4,06 ± 0,01	PE	at T = 174°C (liquid state)	1976	4.60
In	4,09 ± 0,01	PE	at T = 130°C, polycrystalline	1976	4.60
Ir(111) (110)	5,76 ± 0,04 5,42 ± 0,02	FE	--	1973	4.91
K	2,55 ± 0,05* 2,15 ± 0,05** 2,27 ± 0,05***	FE	films in the high coverage limit for substrates *:W(110); **:W(111) ***:W(112)	1975	4.110
	2,30 ± 0,02	PE		1972	4.133
K	2,26* 2,01**	CPD	films in the high coverage limit for substrates: *:Ta(110); **:W(100); reference values: $\Phi_{Ta}(110) = 4,94$ eV, $\Phi_W(100) = 4,65$ eV are assumed.	1971	4.111

Table 4.3 (continued)

Element	Work function [eV]	Technique	remarks	Year	Reference
K	$2,28 \pm 0,05$	PE	T = 25°C	1974	4.51
La	$3,5 \pm 0,2$	PE	polycrystalline film	1970	4.83
Li	2,93	FE	films (d ≈ 1,3 M.L.) evaporated on W(111)/deduced from fig. (1) in reference /4.113/	1968	4.113
Mg	3,66	PE	films on quartz	1964	4.119
Mo	$4,6 \pm 0,15$	PE	polycrystalline film	1970	4.83
Mo(100)	$4,53 \pm 0,02$	PE	- -	1974	4.77
(110)	$4,95 \pm 0,02$				
(111)	$4,55 \pm 0,02$				
(112)	$4,36 \pm 0,03$				
(114)	$4,50 \pm 0,04$				
(332)	$4,55 \pm 0,02$				
Mo(211)	4,58	FE	relative to the average WF of the field emitter	1974	4.78
(100)	4,45				
(111)	4,19				
(321)	4,14				
(310)	4,13				
Mn	$4,1 \pm 0,2$	PE	polycrystalline film	1970	4.83
Na	2,46	CPD	films in the high coverage limit for substrates: (Ta(1ī0); Φ_{Ta}° (110) = 4,94 eV assumed)	1971	4.111

Element	Work function[eV]	Technique	remarks	Year	Reference
Na	2,38			1971	4.107
Na	2,36 ± 0,02	PE	-	1972	4.133
Nb	4,3 ± 0,15	PE	polycrystalline film	1970	4.83
Nb(110)	4,87 ± 0,07	TE	--	1974	4.90
(111)	4,36 ± 0,06				
(112)	4,63 ± 0,06				
(113)	4,29 ± 0,06				
(116)	3,95 ± 0,06				
(001)	4,02 ± 0,06				
(310)	4,18 ± 0,05				
Nb(100)	4,18 ± 0,02	FE	--	1973	4.91
Nd	3,2 ± 0,25	PE	polycrystalline film	1970	4.83
Ni	5,15 ± 0,1	PE	polycrystalline film	1970	4.83
Ni(100)	5,53 ± 0,05	FE	--	1973	4.91
Ni(110)	5,04 ± 0,02	PE	in addition to these bulk results	1971	4.79
(100)	5,22 ± 0,04		measurements are obtained from Ni-		
(111)	5,35 ± 0,05		films epitaxed on mica and rock salt		
Ni(111)	5,42 ± 0,04	PE	--	1969	4.81
Ni(110)	4,64 ± 0,03	TE	Richardson plots were taken in	1969	4.89
(100)	4,89 ± 0,03		the temp. range 1250 - 1570 K		
(111)	5,22 ± 0,03				
Os	5,93	PE	film on W	1962	4.120

Table 4.3 (continued)

Element	Work function[eV]	Technique	remarks	Year	Reference
Pb	4,25	PE	polycrystalline film	1970	4.83
Pd	5,55 ± 0,1	PE	polycrystalline film	1970	4.83
Pd	5,22	PE	equilibrated films (80-120 Å)	1974	4.96
Pt(111)	5,93	FE	reference to total	1973	4.92
(100)	5,84		emission WF (Φ_{tot} = 5,32 eV)		
(331)	5,12				
(320)	5,22				
Pt	5,64	PE	polycrystalline film	1975	4.93
Pt	5,5 ± 0,1	PE	measure performed on a foil	1976	4.94
Pt	5,65 ± 0,1	PE	polycrystalline film	1970	4.83
Rb	2,16 < 0,05	PE	T = 25°C	1974	4.51
Rb	2,05	PE	film evaporated on W(100); Φ final obtained at Θ = 1	1974	4.112
Rb	2,261 ± 0,015	PE	films deposited on quartz; measurement at 140 K.	1974	4.98
Re	4,72	TE and (pos. ion emiss.)	temp. range 1700 K to 2150 K. The WF data, indicated in column 2 is deduced from the Richardson-plot (Re) in ref. /4.116/	1973	4.116
Ru	4,71	PE	at 293 K	1974	4.130
Sb amorph	4,55	-	-	1972	4.131
Sb(100)	4,7	-	-	1971	4.132
Sc	3,5 ± 0,15	PE	polycrystalline film	1970	4.83
Sm	2,7 ± 0,3	PE	polycrystalline film	1970	4.83

Element	Work function [eV]	Technique remarks		Year	Reference
Sn	4,42	CPD	from "metal/insul./metal" - junction experiment; reference: $\Phi_{Al} = 4,08$ eV	1963	4.122
Ta(110)	$\approx 4,8 + 0,6 \times 10^{-4}$ T	TE	after: Table 4 in G.A. HAAS and R.E. THOMAS /4.95/	1966	4.103
(100)	$\approx 4,15 + 2 \times 10^{-4}$ T				
(111)	$\approx 4,00 + 2,6 \times 10^{-4}$ T				
Tc	4,88 ± 0,05	*)	results of electronegativity and the exchange current for the electrolytic hydrogen evolution are used to obtain estimation of this WF data	1972	4.115
Ti	4,33 ± 0,1	PE	polycrystalline film	1970	4.83
Ti	4,2 ± 5,6	CPD	Ti-films evaporated on gold resp. silver (films) in situ; references: $\Phi_{Au} = 5,3$ eV and $\Phi_{Ag} = 4,3$ eV	1973	4.100
V	4,3 ± 0,1	PE	polycrystalline film	1970	4.83
W(poly)	4,6	PE	--	1974	4.80
W(001)	4,60 ± 0,05	CPD	different crystal planes between the <001> to <112> direction are measured ; a value of $4,54_5$ eV was recommended for poly-crystalline W.	1969	4.85
(119)	4,56 ± 0,05				
(116)	4,36 ± 0,05				
(115)	4,35 ± 0,05				
(229)	4,34 ± 0,05				
(114)	4,40 ± 0,05				
(227)	4,43 ± 0,05				
(113)	4,55 ± 0,05				
(112)	4,71 ± 0,05				
(025)	4,55 ± 0,05				

Table 4.3 (continued)

Element	Work function [eV]	Technique	remarks	Year	Reference
W(110)	5,22	TE	at T = 2300 K	1976	4.22
W(111)	4,41	FE	--	1971	4.87
(112)	5,01				
(123)	4,58				
(122)	4,47				
(012)	4,46				
(124)	4,33				
W(001)	4,93 ± 0,06	FE	--	1973	4.88
(111)	4,45 ± 0,05				
(112)	5,12 ± 0,07				
(113)	4,46 ± 0,05				
(013)	4,34 ± 0,04				
(123)	4,50 ± 0,06				
(133)	4,68 ± 0,07				
(233)	4,41 ± 0,05				
(023)	4,58 ± 0,06				
(334)	4,62 ± 0,04				
(223)	4,70 ± 0,05				
(012)	4,36 ± 0,04				
(122)	4,30 ± 0,04				
(124)	4,35 ± 0,05				
(134)	4,74 ± 0,07				
(114)	4,42 ± 0,05				
(116)	4,32 ± 0,06				
(016)	4,43 ± 0,04				

Remarks	Work function [eV]	Technique	remarks	Year	Reference
W(100)	4,63 ± 0,02	FE	--	1973	4.91
(111)	4,47 ± 0,02				
(110)	5,25 ± 0,02				
W(100)	4,57 ± 0,14	FE	--	1975	4.97
(112)	4,89 ± 0,15				
(013)	5,19 ± 0,16				
(111)	4,83 ± 0,38				
Y	3,1 ± 0,15	PE	polycrystalline film	1970	4.83
Yb	2,6 ± 0,05	FE	thick film evaporated on tungsten	1970	4.102
Zn	3,63	PE	the photoemission from the basal plane of single crystals of high purity (99,999%) is substrate	1969	4.104
Zr	4,0	TE	deduced from data: mean electronic WF ϕ^* vs nonmetal atomic content	1972	4.101
Zr	4,05 ± 0,1	PE	polycrystalline film	1970	4.83

* Pauling's relation between electronegativity and WF is used.

of coverage Θ (in fractional monolayers).[1] ϕ and ϕ_s are the WF of the covered and
the clean surface respectively. Generally the function $\Delta\phi = f(\Theta)$ varies considerably
both in sign and in curve shape even for a specific system since it is dependent on
many secondary parameters such as adsorption-constituents, temperature, and pressure,
as well as on kind of crystal face, smoothness etc. as will be shown in Section (5.2).

For coverages of only a few tenths of a monolayer (low coverage limit) $\Delta\phi$ is frequently observed to be proportional to Θ. Bearing in mind these facts it is sensible
to regard WF experiments as falling into two basic categories, viz.

(I) $\Delta\phi$ is used mainly as a measure of coverage;

(II) WF data are used to evaluate representative parameters of the
      adsorbate system, like dipole moments, the arrangement of the
      adatoms, etc.

In both cases suitable coexperiments have to be performed to ensure calibration and
additional in situ experiments (e.g., LEED, Auger, Photoemission, etc.) are necessary
to obtain as complete information as possible.

Chapter 5 is thus divided in three sections. Section 5.1 deals with the possibility
of measuring fractions of a monolayer in the limit of small coverages. Section 5.2
is devoted to WF variations probing adsorbate induced changes of the electronic
structure of metal surfaces. The respective results for a large number of ad-systems
are finally compiled in Sect. 5.3.

5.1 Work Function as a Measure of Coverage

If we consider small coverages without any mutual interactions between the adsorbated
particles, the elementary theory yields /5.2/

$$\Delta\phi \sim \Theta .$$
(5.1)

5.1.1 Calibration Methods

Many attempts have been made to obtain a calibration of $\Delta\phi$ against coverage. The
most accurate calibration would be an absolute determination of the coverage by
means of gravimetric or capillary flux methods in a glass apparatus. Using stainless
steel systems with their large effective area, it is much more convenient to monitor
either Auger spectra or flash desorption traces combined with LEED observations for

---

[1] In the literature, both the number N_a of adsorbed particles per unit area and the
number of fractional monolayers $\Theta = N_a/\bar{N}_a$ of the adsorbate are used where $\bar{N}_a$ is
the number of adatoms in a complete monolayer.

an independent calibration. Molecular beam studies can also be used for this purpose
as was shown recently by ENGEL and ERTL /5.3/ for the system CO/Pd(111). CHRISTMANN
et al. /5.4/ used flash desorption spectra in the case of H_2 adsorption on Ni single
crystal surfaces, and obtained a set of curves corresponding to various coverages
of H_2 on Ni(110) in the range of 0.17...1.55 Langmuir, shown in Fig. 5.1. To check

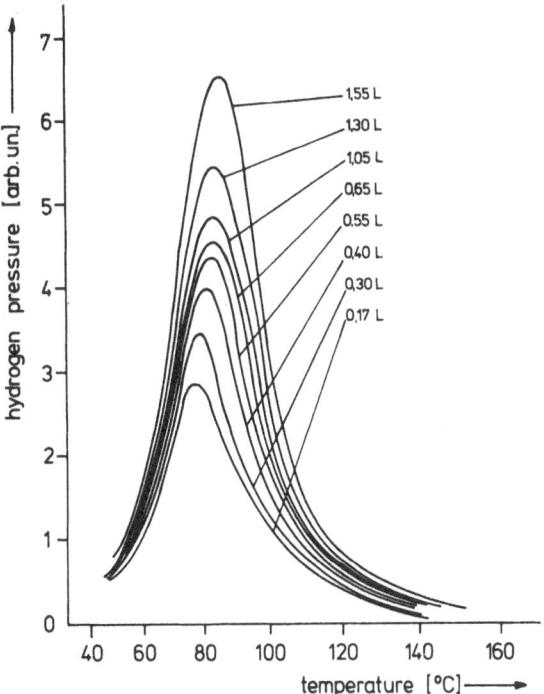

Fig. 5.1. Flash desorption spectra for the initial stages of H_2 adsorption on
Ni(110). After CHRISTMANN et al. /5.4/

the proportionality between coverage and WF variation, the authors plotted the
area underneath each desorption curve, $\int pdt$, for a certain coverage, against the
WF measured at the same coverage. The result is shown in Fig. 5.2. In a more recent
work CHRISTMANN et al. /5.5/ have performed analogous calibration procedures for
the system H_2/Pt(111) in a more subtle manner. Particular attention was paid to
inaccuracies in measuring the absolute number of particles being desorbed in the
flash experiment. They found that $\Delta\Phi$ - mainly in the range of extremely small
coverages - appeared to be strongly dependent on geometrical disordering of the

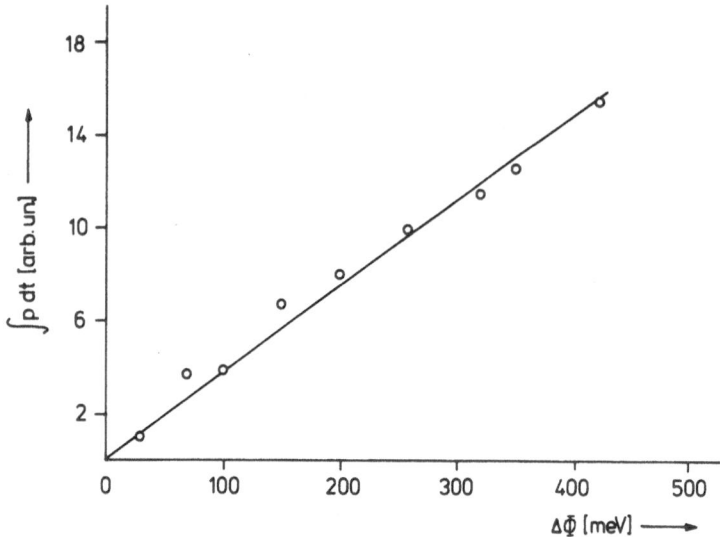

Fig. 5.2. Calibration of WF change $\Delta\Phi$ for $H_2/Ni(110)$ against the relative adsorbed amount as determined from the areas $\int p \, dt$ below the corresponding flash desorption curves. After CHRISTMANN et al. /5.4/

monocrystalline surface in question. Deviations from direct proportionality were found with this system; the authors observed an over-proportional increase of $\Delta\Phi$ with Θ obeying the relation

$$\Delta\Phi = -0,23 \; \Theta^{1,33} \; \text{[eV]} \; , \tag{5.1a}$$

which was ascribed tentatively to long-range interactions between the adsorbed particles even in the low coverage stage. It is worth noting that within some restrictions (e.g., not suitable for hydrogen adsorption), calibration can be performed conveniently by means of Auger electron spectroscopy /5.6/. Particular care has to be taken when using such a procedure since possible electron stimulated effects (desorption or dissociation) may produce incorrect results.

5.1.2 Use of the Work Function as a Measure of Coverage

Many investigations which needed a knowledge of coverage within the low coverage limit, have been performed by observing only $\Delta\Phi$, the variation of WF, which can be measured in a simple manner and with high accuracy.

Thermodynamics

As a representative example within this category, the evaluation of adsorption isotherms and isothermic heats of adsorption E_{ad} as a function of Θ should be mentioned,
as these important thermodynamic properties have been derived for many adsorption
systems on the basis of WF data /5.7/. In these studies it is assumed that $\Delta\Phi$ is a
unique function of Θ and does not depend on temperature. For an overlayer in equilibrium with the gas phase the Clausius-Clapeyron equation holds

$$\left. \frac{d \ln p}{d (1/T)} \right|_{\Theta=const.} = - \frac{E_{ad}}{R} ,$$

(5.2)

where p and T, respectively, represent gas pressure, and absolute temperature of the

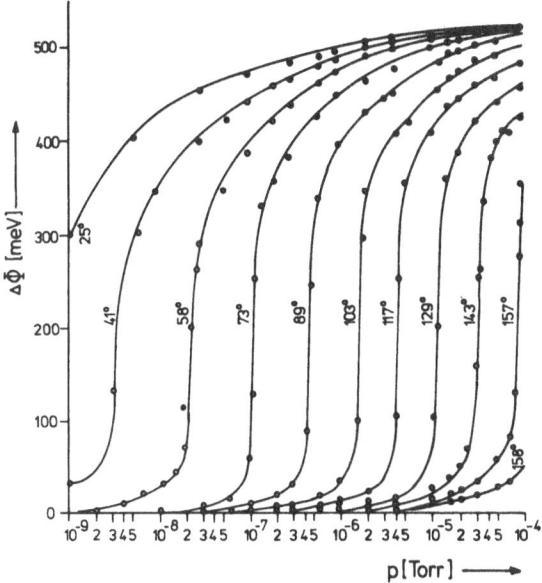

Fig. 5.3. Adsorption isotherms $\Delta\Phi$ vs. pressure of $H_2/Ni(110)$. After CHRISTMANN
et al. /5.4/

adsorbate. R is the gas constant. In Figs. 5.3,4 examples are shown from the
measurements of CHRISTMANN et al. /5.4/ on the system $H_2/Ni(110)$.

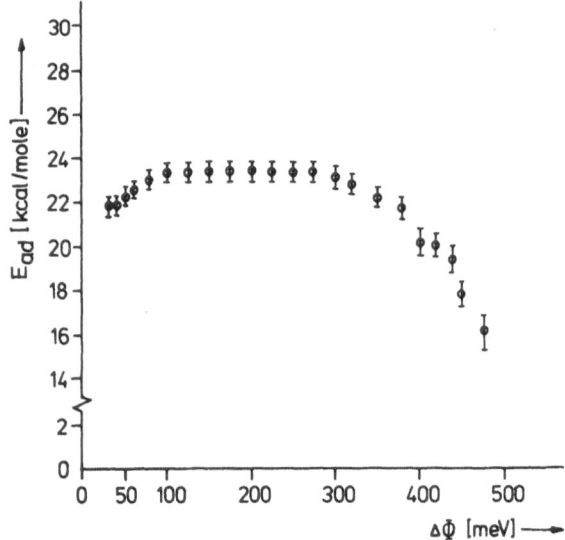

Fig. 5.4. Isosteric heats of H_2 adsorption on Ni(100) as a function of $\Delta\Phi$. After CHRISTMANN et al. /5.4/

Surface Kinetics

In addition to the thermodynamic properties of an adsorption system, information about the surface kinetics is desirable. In the present context the kinetics of adsorption can be studied by monitoring WF variations as a function of time. Thus the sticking coefficient function

$$s = s_0 f(\Theta),\qquad(5.3)$$

where s_0 is the sticking coefficient at zero coverage, can be determined as follows. Simultaneous adsorption and desorption from a surface with nonactivated adsorption can be described by the rate equation

$$\frac{dN_a}{dt} = s_0 f(\Theta)\dot{N} - \delta g(\Theta),\qquad(5.4)$$

where N_a equals the number of adsorbed particles, and the first term on the right hand side represents the rate of adsorption, with $\dot{N}$ being the number of particles per second impinging from the gas phase (pressure p) at a temperature T onto the surface. The second term is the rate of desorption where δ is the rate constant for desorption and $g(\theta)$ contains the over-all order of the desorption process. Consequently dN_a/dt represents the time derivative of the number of particles actually being adsorbed on the surface. $\dot{N}$ can be obtained from the kinetic theory of gases

$$\dot{N} = p(2mk_BT)^{-1/2}, \qquad\qquad (5.5)$$

where m is the mass of the particle and k_B Boltzmann's constant. Now dN_a/dt is related closely to the variation of WF with time.

In the case of $\delta \approx 0$ (desorption neglected) the dependence $s = s(\theta)$ can therefore be evaluated very easily by following $\Delta\phi(t)$. If $\delta \neq 0$ an apparent sticking coefficient s_{app} can be obtained.

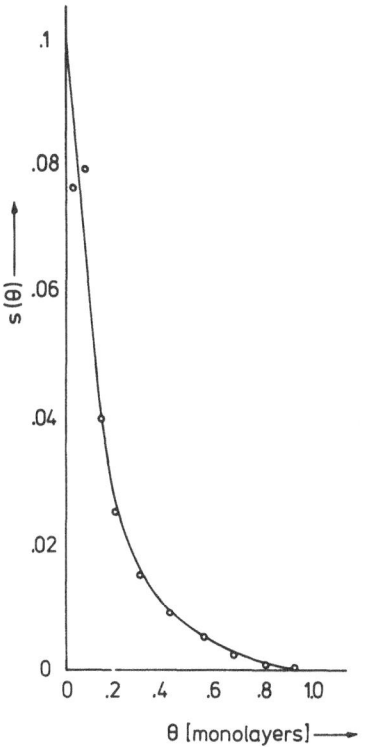

Fig. 5.5. Sticking coefficient of $H_2/Pt(111)$; $s(\theta)$ versus θ. After CHRISTMANN et al. /5.5/

As an example in Fig. 5.5 the sticking coefficient $s(\theta)$ for $H_2/Pt(111)$ is shown plotted against θ as obtained by CHRISTMANN et al. /5.5/ on the basis of WF data.

Surface Diffusion

As a final example within this section it will be demonstrated to what extent a
suitable WF experiment can be used to evaluate surface diffusion parameters of gas
covered macroscopic substrates.

Using microscopic systems GOMER and his coworkers /5.8-10/ have studied such
phenomena using mainly field electron microscopic (FEM) techniques. They have obtained some valuable results in particular with respect to the moving of adsorbate
boundaries, to local adsorption, and to gradual changes of the adsorbate. In addition
BUTZ and WAGNER /5.11/ have investigated the macroscopic system O/W(110) in the
temperature range 760 - 880°C up to a coverage of one monolayer. In their experiment
the authors proceeded as follows. After having carefully cleaned the probe, one half
of the monocrystalline sample surface was covered very closely with a shield and the
open surface area was exposed to oxygen to a given coverage Θ. Starting with such
a step-shaped adsorbate/substrate arrangement, the sample was then heated to a certain preset elevated temperature T for a fixed diffusion time t_D (t_D was chosen in
the range 5, ... 45 min). After the time $t = t_D$ was passed the oxygen concentration
c(x) along one direction was monitored in order to evaluate the diffusion coefficient
D, which is known to be dependent on c, that is $D = D(c)$. To obtain D(c) Fick's
second law

$$\frac{\partial c}{\partial t} = \frac{\partial}{\partial x} \left(D(c) \frac{\partial c}{\partial x} \right) \tag{5.6}$$

with c the number of adsorbed atoms per cm^2, was used. Very simple mathematical trans-
formations as applied by MATANO /5.12/ then yielded

$$D(c) = -\frac{1}{2t} \frac{\int_{c_0}^{c} x\,dc'}{dc/dx} \tag{5.7}$$

with $\int_{c_0}^{c_1} x\,dc' = 0$

where $c = c_0$ for x>o; t=o

$\qquad\qquad$ x→∞; t>o

$\qquad c = c_1$ for x<o; t=o

$\qquad\qquad$ x→-∞; t>o .

From this, knowing $c = c(x)$ for a given T and t_D, the diffusion coefficient D(c)
could be calculated. At that point the correlation between the gas coverage Θ of

the substrate and its change in WF was again used. To do this, a planar calibration experiment as mentioned above was performed, in which the oxygen AES signal heights were correlated with $\Delta\Phi$ the WF variation which, of course, had to be measured with sufficient lateral resolution. Thus, the concentration profile $c(x)$, was obtained. A sophisticated device for measuring $\Delta\Phi$ with high lateral resolution has been developed by the same authors as described in Sect. 3.3.2d.

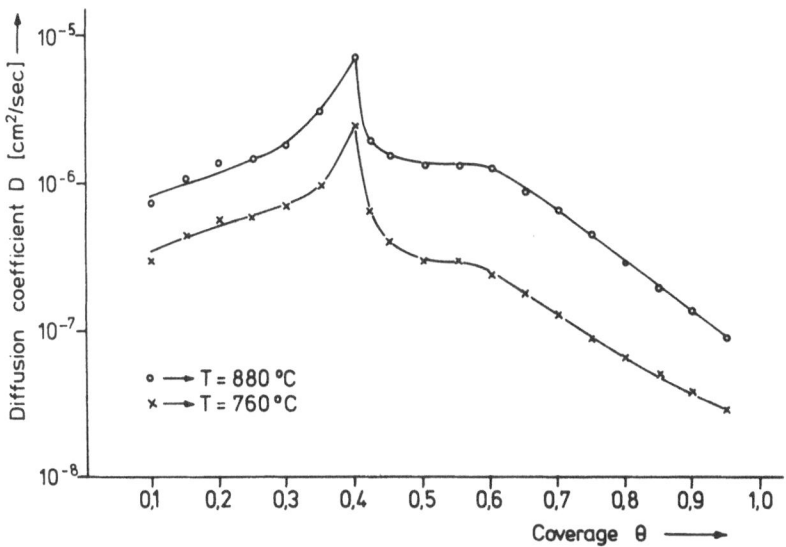

Fig. 5.6. Diffusion coefficient D as a function of coverage Θ for the system O/W(110) for two different temperatures. After BUTZ and WAGNER /5.11/

In this way BUTZ and WAGNER obtained the diffusion coefficient D as a function of coverage for T = 880°C and T = 760°C, shown plotted in Fig. 5.6. The reader interested in the discussion of results is referred to the original paper.

Up to now WF experiments have been utilized for an evaluation of the number of gaseous particles adsorbed on metal substrates, from which information on valuable thermodynamic and kinetic properties of the system in question can be obtained.

Experiments of this kind may be extended to an analysis of systems where a metal vapor is condensed onto a metal substrate, as was shown recently by FEDORUS et al. /5.13/ and by VEDULA et al. /5.14/. The second category of WF experiments mentioned in the Introduction to this chapter tends to correlate more or less basic theoretical models of the WF with electronic surface behavior, in order to evaluate representative surface properties in detail. This category will be discussed in Sect. 5.2.

5.2 Work Function and Representative Surface Data of Adsorbate Systems

In addition to the use of the WF as a measure of coverage (see Sect. 5.1), there are
numerous experiments, which aim at characterizing the ad-system in terms of representative electronic and structural surface parameters (e.g., charge of adatom,
dipole moment, rearrangement quantities, etc.). This can only be achieved within
the framework of suitable models, which can be divided into two basic categories
according to the behavior of the substrate:

    (I)   Static substrate model

          The structure of the substrate remains unchanged during

          an adsorption process.

    (II) Dynamic substrate model

          The structure of the substrate surface changes during

          the adsorption process.

    Most adsorption studies have been based on the static substrate model (I); various
classical and quantum-mechanical approaches allowing for adsorbate/substrate and
adsorbate/adsorbate interaction mechanism have been made during its application.
(See Sect. 2.3).

    Investigations based on the dynamic substrate model (II) (i.e., treating recon-
struction phenomena, etc.) are quite new and promise further insight into the electronic behavior of solid surfaces. A few experiments and attempts at interpretation
will be reviewed below, (see Sect. 5.2.2). The following section is devoted to the
discussion of the static substrate model.

5.2.1 Static Substrate Model

a) Theoretical Relationship and Basic Adsorption Experiments

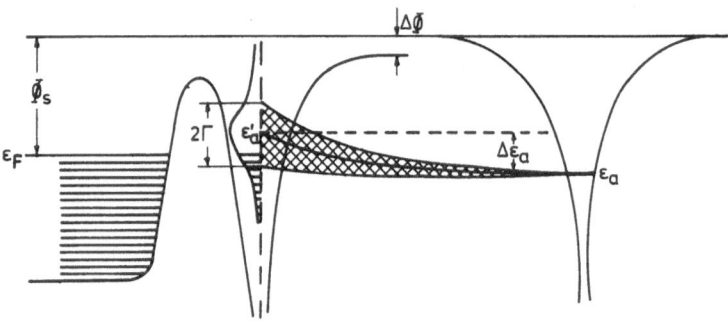

Fig. 5.7. Energy level diagram relevant to a metal surface with an adsorbed alkali
atom. Symbols are explained in the text. Taken from /5.18/

Without going in detail into theoretical treatments (given in Sect. 2.3) for the purposes of a more coherent discussion of the main experiments covered in this topic, it seems to be useful to sketch the essential points of GURNEY's model /5.15/, which can be regarded as one of the starting points in the explanation of the various adsorption phenomena. According to this model (see Fig. 5.7) an energy level ε_a of an atom approaching a metal surface is shifted by an energy $\Delta\varepsilon_a$ and at the same time broadened to a width of half height 2Γ when the adatom approaches its equilibrium position near the substrate surface. In Fig. 5.7 the change of the valence level of an alkali atom is shown qualitatively.

Quantitatively the local density of states $n_a(\varepsilon)$ of an electron on the adatom can be described by a Lorentzian

$$n_a(\varepsilon) = \frac{\Gamma}{(\varepsilon - \varepsilon_a')^2 + \Gamma^2} \ , \qquad (5.8)$$

with $\quad \varepsilon_a' = \varepsilon_a + \Delta\varepsilon_a$.

The energy levels are filled up to the Fermi level ε_F (at zero temperature) such that there are

$$<n_a> = \int_{-\infty}^{\varepsilon_F} n_a(\varepsilon) \, d\varepsilon \qquad (5.9)$$

electrons in the considered state of the adatom. The positive charge of the adatom is therefore given by

$$q = e \ (1 - <n_a>) \qquad (5.9a)$$

(typically ≈ 0.5 e for alkali/transition metals /5.16/; e denoting the elementary charge). In the simplest case (i.e., for a jellium structured substrate) the negative of the charge on the adatom is localized at the metal surface in the form of a charge density which screens the field of the charged adatom. This screening charge and the charge q on the adatom form a dipole $p = q \cdot a$, where a denotes the separation between the adatom and the center of gravity of the screening charge. For the simple classical image-charge concept 2a is the distance between the charged adatom and its image. The WF change due to N_a per unit area is given by Helmholtz equation[2]

$$\Delta\Phi = - \ 4\pi e p N_a \qquad (5.10)$$

---

[2] Sometimes the definition of the dipole moment differs by a factor 2 from that given here, so that the Helmholtz equation contains a factor 2 instead of 4. Φ in eV, p in Debye and N_a in adatoms/cm^2 are interrelated by the equation [using the factor 4 in (5.10)]

$$\Delta\Phi = - \ 3{,}7673 \cdot 10^{-15} \cdot p \cdot N_a \ .$$

where $\Delta\Phi$ is seen to be linearly dependent on N_a (low coverage case; p is assumed to be independent of N_a). In this case only adsorbate/substrate interaction is considered. With the use of (5.10) the initial dipole moment p_0 can be obtained directly from elementary experiments in which the dependence of $\Delta\Phi$ on N_a is studied

$$p_0 = - \frac{1}{4\pi e} \left(\frac{\partial \Delta\Phi}{\partial N_a}\right)_{N_a \to 0} . \qquad (5.10a)$$

The calculation of p_0, carried out with the use of classical and older quantum-mechanical models up to now, has been improved considerably by means of the so-called semiclassical and the more refined quantum-mechanical approaches (see e.g. GYFTOPOULOS-STEINER /5.17/, GADZUK /5.18/, NEWNS-ANDERSON /5.19/, etc.). Table 5.1 shows p_0 values deduced from various adsorption measurements compared with the results of some of the above mentioned theoretical treatments.

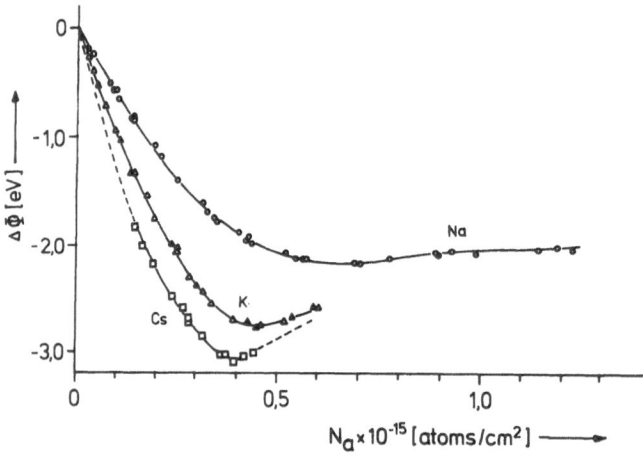

Fig. 5.8. WF change $\Delta\Phi$ as a function of coverage when Na, K and Cs are evaporated onto Ni(100). After GERLACH and RHODIN /5.20/

In all the experiments higher coverages ($\Theta \lesssim 1$) were also studied, and hence not only adsorbate/substrate but also adsorbate/adsorbate interaction, i.e., mutual influence of dipoles must be considered. The adsorption of alkali metals on pure metal surfaces represents one of the simplest cases for the study of these various questions. Fig. 5.8 shows the WF variation of Ni(110) as a function of Na, K and Cs coverage according to experiments carried out by GERLACH et al. /5.20/.

Table 5.1

System	2p_o [Debye] (exp.)		2p_o [Debye] (theor.)			According to
	from (5.10a)	from (5.11a)	$2p_o^C$	$2p_o^{G-S}$	$2p_o^{GA}$	
1. Cs/Ta(110)	15.3 ± 0.7	--	16,2	--	17.0	FEHRS and STICKNEY /5.30/ (1971)
2. K/Ta (110)	13.0 ± 0.8	--	12.8	--	12.7	
3. Na/Ta(110)	8.5 ± 0.6	--	9.1	--	7.5	
4. Cs/W (100)	15.0 ± 0.8	--	16.2	--	12.9	
5. K/W (100)	13.5 ± 0.7	--	12.8	--	9.2	
6. Na/Ni(111)	7.4 ± 0.5	--	--	8.0	8.5	GERLACH and RHODIN /5.20/ (1970)
7. Na/Ni(100)	7.2 ± 0.5	--	--	8.0	6.9	
8. Na/Ni(110)	3.2 ± 0.3	--	--	6.8	4.2	
9. K/Ni (110)	5.3 ± 0.5	--	--	8.9	7.2	
10. Cs/Ni(110)	7.0 ± 0.7	--	--	10.6	11.0	
11. K/W(110)	15.7	15.8	--	--	--	SCHMIDT and GOMER /5.22/ (1966)
12. K/W(211)	13.6	16.0	--	--	--	
13. K/W(111)	9.5	↑0.0	--	--	--	
14. K/W(100)	11.5	↑1.5	--	--	--	
15. Na/Pt(111)	10.5	10.4	9.1	8.0	--	MEISTER, MALZFELDT, HÖLZL /5.28/ (1978)
16. Na/Ni(100)	6.8	--	--	--	--	ANDERSSON and JOSTELL /5.16/ (1974)
17. K/Ni (100)	13.5	--	--	--	--	

Two unique characteristics can be deduced from these experiments and from analogous ones by other workers:

(I) A nearly linear decrease of $\Delta\Phi$ in the low coverage limit, reflecting almost pure adsorbate/substrate interaction.

(II) A pronounced deviation from linearity increasing coverage as a consequence of the mutual influence of dipoles.

(Note that, with further increase of coverage, in most cases (but not all) a minimum value $\Delta\Phi_m$ is observed; finally a value $\Delta\Phi_a$ is obtained corresponding to the bulk WF Φ_a of the adsorbate).

Treating the latter behavior (II), TOPPING /5.21/ has presented a very fruitful idea on a simple point depolarisation model which yields

$$\Delta\Phi = -\frac{4\pi e p_0 N_a}{1+9\alpha N_a^{3/2}} , \qquad (5.11)$$

where p_0 is the initial dipole moment and α denotes an effective polarisibility. On inserting $\Theta = N_a/\bar{N}_a$ into (5.11), where $\bar{N}_a$ is the number of adatoms in a complete monolayer, and rearranging (5.11), one obtains

$$\Theta/\Phi = (1/C_1) + (C_2/C_1) \Theta^{3/2} \qquad (5.11a)$$

where

$$C_1 = (\partial\Delta\Phi/\partial\Theta)_{\Theta\to0}$$

$$C_2 = 9\alpha\bar{N}_a^{3/2} .$$

If $\Theta/\Delta\Phi$ is plotted vs. $\Theta^{3/2}$ a straight line should result for a given adsorbate/ substrate system, from which both the initial dipole moment p_0 and the polarisability α can be determined. SCHMIDT-GOMER /5.22/ have checked this relation by studying the adsorption of K on various surface planes of W. The result is shown in Fig. 5.9. In a more recent study of Na and K adsorbed on Ni(100) ANDERSSON and JOSTELL /5.16/ arrived at results which agreed equally well with TOPPING's theory.

In Table 5.1 various results have been collected all of which deal with the initial dipole moment p_0. In the first group of results (rows 1 to 10) experimental values of $2p_0$ are compared with values derived from various theoretical models. Surprisingly, the values obtained for p_0^C based upon the classical image model agree very well with the experimental data. By contrast, the values for p_0^{G-S} and p_0^{GA} derived from the more refined theories of GYFTOPOULOS-STEINER /5.17/ and of GADZUK /5.18/ respectively, do not appear to agree so satisfactorily.

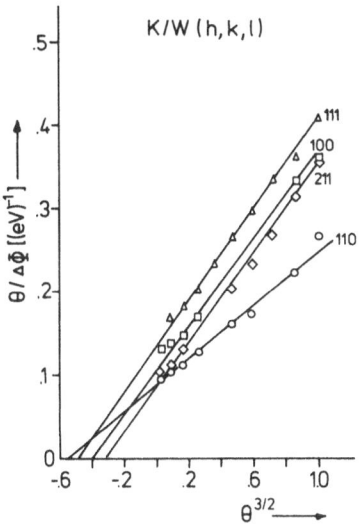

K/W (h,k,l)

Fig. 5.9. Plots of $\Theta/\Delta\Phi$ versus $\Theta^{3/2}$ for K on various planes (hkl) of W. After SCHMIDT and GOMER /5.22/

In the second group (rows 11 to 14) the above mentioned comparison of results obtained on the basis of (5.10a) and TOPPING's equation (5.11a) is shown. In most cases there is a good agreement, apparantly confirming the validity of TOPPING's idea. Rows 15 to 17 show a few more recent experimental results.

b) **Experiments Relating to Gurney's "Depolarization Model"**

In the so-called TOPPING model the physical origin of the depolarization is not clear. Gurney gave a qualitative explanation for this phenomenon. According to his model the energy ε_a' (see Fig. 5.7) is pushed down by Coulomb interactions with all other charged adatoms and with their screening charges. Thus, via (5.9), the charge q on the adatom, and thence the dipole moment p, which is proportional to q, is diminished.

In order to examine these more refined theoretical predictions, suitable experiments were necessary. In this connection PLUMMER and YOUNG /5.23/ and GADZUK /5.24/ used resonance tunelling in field emission, and HAGSTRUM and BECKER /5.25/ applied ion neutralization spectroscopy. In recent experiments characteristic energy loss spectroscopy has been used by ANDERSSON and JOSTELL /5.16/ to determine $\Delta\varepsilon_a$ and Γ. As the latter procedure is closely related to the measurement of $\Delta\Phi(\Theta)$ the main principles of this treatment will now be sketched briefly.

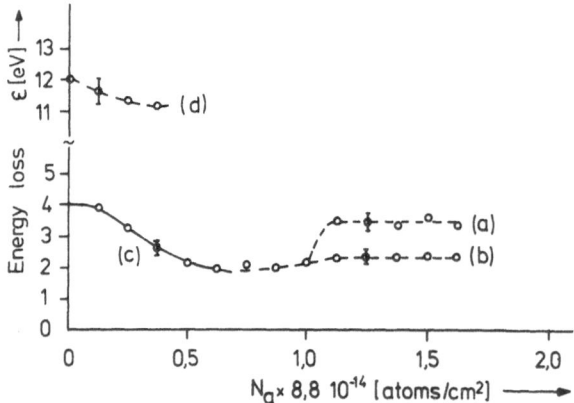

Fig. 5.10. Various (a,b,c,d) energy losses ε vs. coverage for the system Na/Pt(111) as measured by SCHRÖDER and HÖLZL /5.26/ (Primary energy E_p = 30 eV). The solid line passing through points "C" is discussed in the text. The curves a, b and d are not discussed here

In Fig. 5.10 characteristic energy losses (a,b,c and d) obtained recently by SCHRÖDER and HÖLZL 5.26/ are presented. They investigated the system Na/Pt(111) using a primary energy E_p = 30 eV at T = 170 K.

Only loss "c" within the coverage range $0 < N_a < 6,2 \cdot 10^{14}$ atoms cm^{-2} is considered here. Following the treatment of ANDERSSON and JOSTELL /5.16/, this energy loss can be interpreted in terms of an intermetallic transition from the Pt valence band to a virtual 3s level of the Na adatom. Consequently, the coverage dependence of the energy loss $\varepsilon(N_a)$ is related to the coverage dependence of $\varepsilon_a'(N_a)$ which, when introducing $\varepsilon_a'(0)$ as the zero contribution, is given by

$$\varepsilon_a'(N_a) = \varepsilon_a'(0) + \Delta\varepsilon_a'(N_a) \tag{5.12}$$

Here $\Delta\varepsilon_a'(N_a)$ is ascribed to the electrostatic potential at an adatom, due to the Coulomb interactions with all other adatom charges q and their images. For the low coverage limit MUSCAT and NEWNS /5.27/ showed that this energy is given by

$$\Delta\varepsilon_a' = -2A_M eap_o N_a^{3/2}, \tag{5.13}$$

where A_M is a constant (Madelung constant), which is slightly dependent on the geometrical structure of the adsorbate layer and which is approximately equal to 9 /5.21/. Using (5.10) we may eliminate p_o from (5.13) and obtain

$$\Delta\varepsilon_a'(N_a) = \frac{1}{2\pi}A_M a \, \Delta\Phi(N_a) \cdot N_a^{1/2}. \tag{5.14}$$

The dependence of the energy loss ε on coverage is then given by

$$\varepsilon(N_a) = \varepsilon(0) + \frac{1}{2\pi M} A_M a \Delta\Phi(N_a) \cdot N_a^{1/2}. \tag{5.15}$$

Using the known dependence $\Delta\Phi(N_a)$ as measured by MEISTER et al. /5.28/, (5.15) can be fitted with the two parameters $\varepsilon(0)$ and a, to the experimental data $\varepsilon(N_a)$ obtained by SCHRÖDER and HÖLZL /5.26/. In the limit of low coverage these authors obtain the best fit, indicated by the solid line of curve "c" (see Fig. 5.10), by using $\varepsilon(0) = 4.0 \pm 0.3$ eV and $a = 1.9 \pm 0.2$ Å.

Once a is known the charge q_0 (in the limit $N_a \to o$) can be obtained from the Helmholtz equation (5.10a).

With the value of a determined above and the initial slope of WF data of MEISTER et al. /5.28/, both the initial dipole moment $2p_0$ and the adatom charge q_0 (and $\bar{q}_0 = e - q_0$) can be obtained, viz.

$2p_0 = (10.5 \pm 0.4)$ Debye
and
$q_0/e = 0.58 \pm 0.08$.

A comparison of the electron energy loss spectrum with the observed XPS spectrum allows also a rough estimate of ε_a' relative to the vacuum level ϕ_{vac} and thereby (see Fig. 5.7) an estimate of $\Delta\varepsilon_a$ /5.26/.

In Table 5.2 results for a few alkali/metal systems are presented.

Table 5.2

	SCHRÖDER-HÖLZL /5.26/[3] Na/Pt(111)	ANDERSSON-JOSTELL /5.16/ Na/Ni(100)	K/Ni(100)
$2p_0$ [Debye]	10.5 ± 0.4	6.8	13.5
a [Å]	1.9 ± 0.2	1.5	2.5
$\varepsilon(0)$ [eV]	4.0 ± 0.3	3.5	3.6
qo/e	0.58 ± 0.08	0.47	0.56

[3] An error in the numerical data from /5.26/ has been eliminated in Table (5.2).

c) Dependence of ΔΦ(Θ) on the Structure of the Substrate Surface

As a consequence of the adsorbate/substrate interaction sketched in Sect. 5.2.1a
the amount of adsorbate charge, dipole moment, etc., is expected to depend not only
on the species of the two constituents but also on the kind of surface structure of
the substrate in question. For example the initial dipole moment p_o of the Na/Ni(100)
system differs considerably from that on the Na/Ni(110) system /5.20/ as can be seen
from Table 5.1.

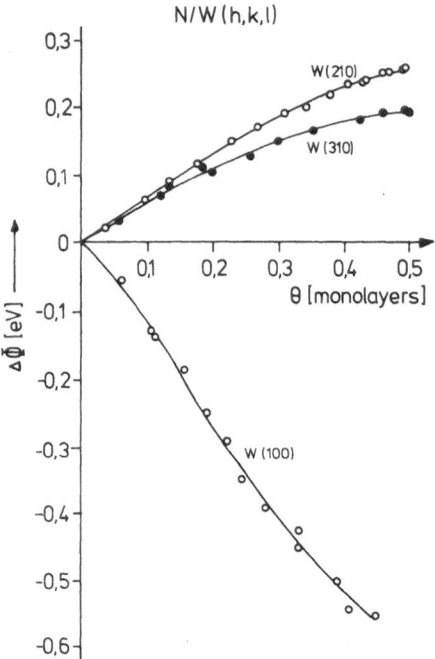

Fig. 5.11. WF change ΔΦ on various single crystal planes (h,k,l) of W, versus nitrogen coverage Θ in monolayers. After ADAMS and GERMER /5.29/

ADAMS and GERMER /5.29/ have obtained a very dramatic result during a gas-adsorp-
tion experiment, the outcome of which is shown in Fig. 5.11. The systems studied by
them were N/W(100), (310) and (210) at 300 K. They observed a distinct decrease of
ΔΦ vs. coverage for the W(100) plane, an increase in the case of the W(210) and W(310)
planes. To explain this the authors propose a simple model which correlates the substrate structure with the change of ΔΦ as a function of coverage. The number of atoms

adsorbed at saturation is assumed to be equal to the number of (100) sites for all planes except those having adjoining sites of this kind in the [001] direction. Additionally it is assumed that for any plane (hk0), the nitrogen atom in a (100) site is always at the same distance above the (100) plane which is associated with the site and that the dipole moment varies linearly with the distance of the nitrogen atom from the (hk0) plane. It is further assumed that the dipole is positive when the nitrogen atom is above the geometric plane, and negative when it is below the geometric plane. The model simulates the experimental data for saturation nitrogen coverage quite well.

5.2.2 Dynamic Substrate Model

Where adsorption experiments are discussed in terms of the static substrate model no variation of crystallographic or geometric ordering of the substrate as a consequence of the adsorption process is assumed. In many cases, however, especially if any heat treatment is applied during the experimental study, a variation of the ordering of the ad-system, dependent both on coverage and temperature is to be expected. It is useful to apply in this case the dynamic substrate model and rearrangement processes have to be discussed.

In this connection BAUER et al. /5.6,31,32/ have recently carried out a number of experiments, in which they measured $\Delta\Phi$ vs. coverage at room temperature for various systems using different deposition temperatures. In Fig. 5.12, one of the main results obtained by BAUER et al. /5.32/ in a study of the system Cu/W(100) is shown. The $\Delta\Phi$ vs. coverage dependence of the adsorbate deposited at room temperature (open circles), is completely different from that of the adsorbate deposited at 800 K (full circles). As can be seen from this figure not only are the initial slopes of the $\Delta\Phi$ vs. coverage plots of opposite sign for the two temperature conditions, but in addition there are maxima and minima at higher coverages.

To explain this BAUER et al. invoked the influence of atomic roughness on the WF (SMOLUCHOWSKI's model /5.33/, see Sect. 2.2.4d). If an atom is adsorbed on a substrate, an additional characteristic variation of WF is to be expected, due to the resulting change of the atomic roughness depending on size and location of the adsorbate. This process occurs with rearrangement of the substrate. During adsorption at 800 K, the surface rearranges and changes its atomic roughness in a fashion different from that at room temperature, which causes consequently different WF changes.

One of the most dramatic examples of surface rearrangement is the surface reconstruction in the system oxygen on W(100), which has been studied in considerable detail by BAUER et al. /5.6/ and by KRAMER and BAUER /5.34/. Upon annealing the substrate, when covered with half a monolayer of oxygen, beyond a certain critical temperature of about 600 K (see Fig. 5.13) a sharp decrease in $\Delta\Phi$ from about +0.6 eV to -0.2 eV was found. This decrease is accompanied by a transition from a p(4x1)

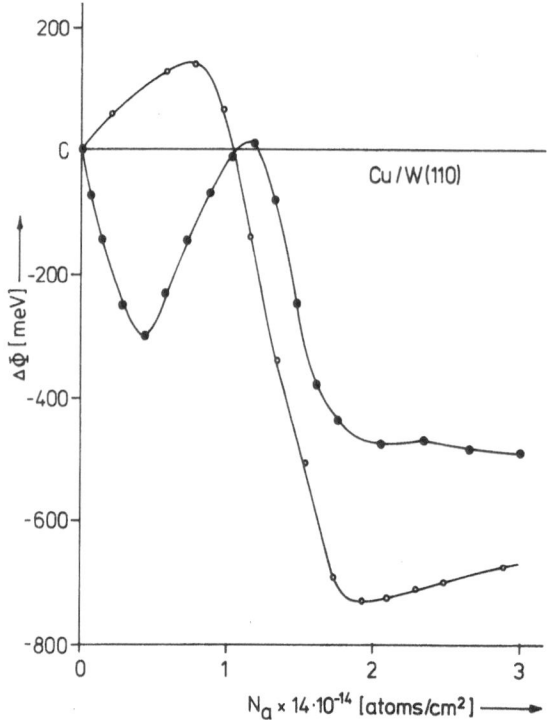

<u>Fig. 5.12.</u> WF change ΔΦ versus coverage in $N_a = 14 \cdot 10^{-14}$ atoms/cm² for deposition at 300 K (open circles) and 800 K (full circles) for Cu/W(110); measurements made at room temperature. After BAUER et al. /5.31/

LEED pattern to a p(2x1) via a(1x1) structure with a diffuse background. The authors interpreted this transition by an exchange mechanism between the W and O atoms, i.e., after annealing half of the atoms in the surface layer are W atoms and the other half are O atoms, the atomic arrangement being such as to yield a p(2x1) structure. Due to the smaller atomic radius of the oxygen atoms (relative to that of the tungsten atoms) this species is visualized as being positioned slightly below the mean W surface thus yielding the negative values of ΔΦ.

In concluding Sect. 5.1,2 , it is obvious that WF variation can contribute in many ways to adsorption studies.

First, as shown in Sect. 5.1, the WF can be used basically as a measure of coverage of the adsorbate atoms where certain surface processes (e.g., thermodynamic, etc.) are studied.

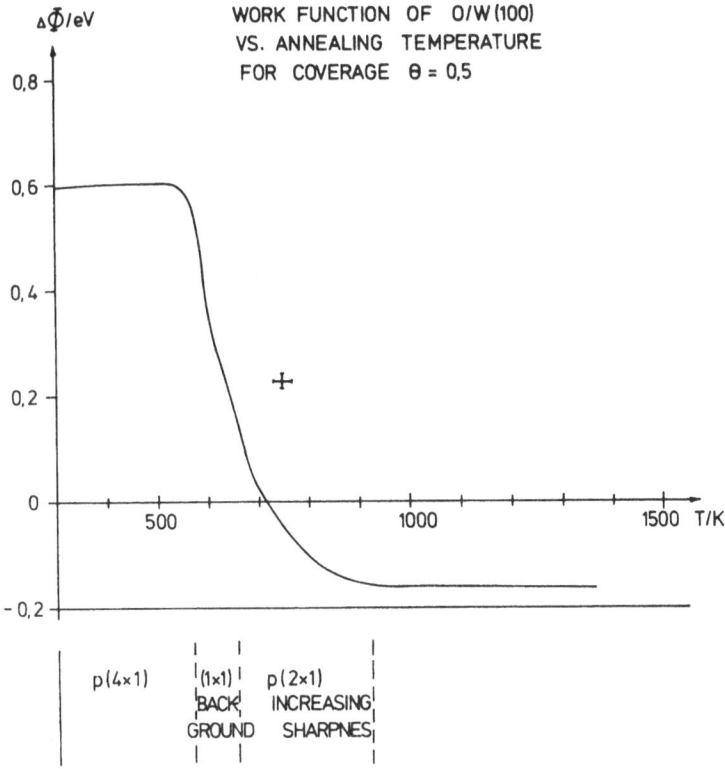

Fig. 5.13. WF Δφ vs. annealing temperature for the system oxygen [0,5 monolayer] deposited at room temperature on W(100), and measured at 300 K. After KRAMER and BAUER /5.34/

Second, as shown in Sect. 5.2, there is a significant possibility that WF results can help to elucidate numerous representative surface parameters.

Due to the fact that experimental facilities in this field have been highly developed in the recent past (e.g., with respect to the variability and flexibility of the experiments), it can be expected that WF results may be enormously helpful in the future for further development of the theory.

5.3 Compilation of Ad-Systems Connected with Work Function Studies

Table 5.3[4] gives an account on ad-systems whose WF has been studied to correlate its variation with surface properties as discussed in Sect. 5.1,2. The aspect of selection employed here is the same as for Table 4.3. Table 5.3 is arranged as follows:

Column 1, the various systems are listed alphabetically with respect to the substrate.

Column 2, the temperature range of measurement is indicated. T_R corresponds to measurements only at room temperature (or lower) and in which no annealing process, etc., has been carried out. T_A corresponds to measurements at room (or elevated) temperature. In addition, in this case, various annealing processes may have been carried out.

Column 3, the sign of the slope of $(\partial\phi/\partial N_a)_{N_a} \to 0$ is indicated. In the case of T_R this sign is either positive or negative. In the case of T_A, both signs (pos. and neg.) are possible.

Column 4, the method of measurement is indicated, viz.
CPD - contact potential difference
PE - photoemission
TE - thermionic emission
FE - field emission

Column 5 and 6, give the year of publication and the reference.

---

[4] The authors are grateful to Dr. A. WACHNIEWSKI for his helpful service in preparing Table 5.3.

Table 5.3

System	Study acc. to: T_R or T_A	Sign (s) of $(\partial\Phi/\partial N_a)_{N_a\to 0}$	Method of measur.	Year	Reference
O/Ag(111)	T_R	pos.	PE	1973	5.55
O/Ag(110)	T_A	pos.	CPD	1976	5.103
O/Ag(111)	T_A	-	CPD	1974	5.131
Xe/Ag	T_A	-	FE	1974	5.139
O/Al	T_A	neg.	CPD	1975	5.35
O/Al	T_R	neg.	CPD	1974	5.38
O/Al	T_A	pos., neg.	CPD	1972	5.85
H_2O/Al	T_A	zero, neg.	CPD	1972	5.90
C_2H_4/Al	T_A	pos., neg.	PE	1975	5.117
C_2H_4/Au	T_A	pos., neg.			
Xe/Au	T_A	-	FE	1974	5.139
NH_3/Au	T_A	pos., neg.	CPD	1975	5.145
CO/Au	T_A	pos., neg.			
wet air/Au	T_A	pos., neg.			
H_2O/Au	T_R	neg.	CPD	1972	5.43
CO/Co	T_R	pos.	PE	1972	5.60
(Cs + O)/Cs	T_R	-	PE	1975	5.42
O/Cu(100)	T_R	pos.	CPD	1971	5.54
/Cu(110)	T_R	pos.			
/Cu(111)	T_R	pos.			
O/Cu	T_R	pos.	PE	1974	5.86
O/Cu(110)	T_R	pos.	CPD	1971	5.66
CO/Cu(100)	T_A	neg.	CPD	1972	5.147
Xe/Cu	T_A	-	CPD	1971	5.129
Xe/Cu	T_A	-	FE	1974	5.139
C_2H_4/Cu	T_A	pos., neg.	PE	1975	5.117
CO/Fe	T_R	pos.	PE	1972	5.60
Xe/Fe	T_A	-	CPD	1971	5.129

Table 5.3 (continued)

System	Study acc. to: T_R or T_A	Sign (s) of $(\partial\phi/\partial N_a)_{N_a\to 0}$	Method of measur.	Year	Reference
Xe/Fe	T_A	-	FE	1974	5.139
Xe/Ir	T_A	-	FE	1974	5.139
CO/Mn	T_R	pos.	PE	1972	5.60
Ag/Mo(100)	T_R	neg.	CPD	1976	5.64
H/Mo(001)	T_R	pos.	FE	1973	5.53
/Mo(011)	T_R	pos.			
/Mo(111)	T_R	pos.			
Ge/Mo	T_R	pos.	FE	1973	5.95
O/Mo(110)	T_A	neg.	CPD	1975	5.58
O/Mo(100		neg.	CPD	1975	5.132
Lu/Mo(110)	T_A	neg.	CPD	1971	5.121
S/Na	T_R	neg.	PE	1971	5.142
Xe/Na	T_A	-	CPD	1971	5.129
O/Nb(110)	T_A	neg.	CPD	1975	5.92
CO/Ni(111)	T_R	pos.	CPD	1974	5.52
CO/Ni(100)	T_A	-	CPD	1972	5.144
CO/Ni	T_R	pos.	CPD	1974	5.84
Cs/Ni(100)	T_R	neg.	CPD	1975	5.79
O/Ni(100)	T_A	-	CPD	1974	5.112
O/Ni(111)	T_A	-	CPD	1974	5.113
O/Ni(100)	T_R	pos.	PE	1976	5.126
O/Ni(100)	T_R	neg.	CPD	1973	5.77
O/Ni(110)	T_R	pos.	CPD	1971	5.66
H/Ni(110)	T_R	pos.	CPD	1973	5.82
H/Ni(110)	T_R	pos.	CPD	1974	5.4
K/Ni	T_R	neg.	CPD	1973	5.40
Na/Ni(111)	T_R	neg.	CPD	1970	5.20
/Ni(100)	T_R	neg.			
/Ni(110)	T_R	neg.			

Table 5.3 (continued)

System	Study acc. to: T_R or T_A	Sign (s) of $(\partial\phi/\partial N_a)_{N_a\to 0}$	Method of measur.	Year	Reference
K/Ni(110)	T_R	neg.			
Cs/Ni(110)	T_R	neg.			
C_2H_4/Ni	T_A	pos., neg.	PE	1975	5.117
C_2H_2/Ni(111)	T_A	neg.	CPD	1974	5.107
C_2H_4/Ni(111)	T_A	neg.			
C_6H_6/Ni(111)	T_A	neg.			
(H,O,C,CO, S,Se,Te)/Ni(001)	T_A	neg., pos.	CPD	1974	5.108
/Ni(110)	T_A	neg., pos.			
/Ni(111)	T_A	neg., pos.			
(O,S,Se)/Ni(111)	T_A	pos., neg.	CPD	1972	5.100
/Ni(110)	T_A	pos.			
/Ni(100)	T_A	pos.			
(Cs+O)/Ni(100)	T_A	neg.	CPD	1975	5.136
Xe/Ni	T_A	–	FE	1974	5.139
Xe/Ni(100)	T_A	pos.	PE	1971	5.150
CO/Pd(110)	T_A		CPD	1974	5.99
/Pd(210)	T_A				
/Pd(311)	T_A				
/Pd(111)	T_A				
H/Pd	T_A	–	CPD	1974	5.149
(H+CO)/Pd(110)	T_A	–	CPD	1974	5.148
C_2H_4/Pd	T_A	pos., neg.	PE	1975	5.117
Xe/Pd(100)	T_R	neg.	CPD	1971	5.80
Xe/Pd	T_A	–	FE	1974	5.139
Ba/Pt	T_A	neg.	FE	1972	5.93
H/Pt(111)	T_A	pos., neg.	CPD	1976	5.5
C_2H_4/Pt	T_A	pos., neg.	PE	1975	5.117
Xe/Pt	T_A	–	FE	1974	5.139
Subsistut./Pt(111) aromatic /	T_R	neg.	CPD	1974	5.115
molecules /Pt(100)	T_R	neg.			

Table 5.3 (continued)

System	Study acc. to: T_R or T_A	Sign (s) of $(\partial\Phi/\partial N_a)_{N_a\to 0}$	Method of measur.	Year	Reference
BaO/Re(0001)	T_A	neg.	CPD	1973	5.59
Ce/Re	T_A	neg.	TE	1975	5.154
O/Re	T_A	pos.	CPD	1972	5.114
Ti/Re	T_R	pos.	FE	1971	5.39
O/Re	T_A	pos.	CPD	1974	5.116
Xe/Rh	T_A	-	FE	1974	5.139
H/Ru	T_A	pos.	FE	1974	5.128
CO/Ru	T_A	pos.	FE	1974	5.127
O/Ru	T_A	pos.			
O/Ru	T_A	pos.	CPD	1975	5.71
Xe/Ru	T_A	-	FE	1974	5.139
Cs/Ta(110)	T_R	neg.	CPD	1971	5.30
K/Ta(110)	T_R	neg.			
Na/Ta(110)	T_R	neg.			
Cl/βTi	T_R	pos.	FE	1971	5.36
Ag/W(110)	T_A	neg.	CPD	1977	5.106
/W(100)	T_A	neg.			
Au/W(110)	T_A	neg.			
/W(100)	T_A	pos., neg.			
Ag/W	T_A	pos.	FE	1974	5.51
Au/W	T_A	pos.			
Cu/W	T_A	pos.			
Au/W	T_R	pos.	FE	1974	5.46
Al/W(001)	T_R	neg.	FE	1974	5.120
/W(111)	T_R	pos.			
Ba/W(110)	T_R	neg.	CPD	1975	5.97
Ba/W(112)	T_R	neg.	CPD	1973	5.72
Ba/W(011)	T_R	neg.	CPD	1972	5.13
Mo/W(011)	T_R	neg.			

Table 5.3 (continued)

System	Study acc. to: T_R or T_A	Sign (s) of $(\partial\phi/\partial N_a)_{N_a\to 0}$	Method of measur.	Year	Reference
Ba/W(100)	T_A	neg.	TE	1974	5.109
/W(110)	T_A	neg.			
/W(111)	T_A	neg.			
(O+Ba)/W(100)	T_A	neg.			
/W(110)	T_A	neg.			
/W(111)	T_A	neg.			
Be/W(211)	T_A	pos., neg.	FE	1976	5.151
/W(111)	T_A	pos., neg.			
/W(110)	T_A	pos., neg.			
Cu/W(110)	T_A	pos., neg.	FE	1973	5.137
/W(100)					
/W(111)					
/W(211)					
Cu/W(211)	T_R	pos.	CPD	1976	5.140
Cu/W(100)	T_A	pos., neg.	CPD	1974	5.32
/W(110)		neg.			
Cl/W(110)	T_R	zero	CPD	1970	5.65
/W(100)		pos.			
/W(111)		pos.			
Br/W(110)		neg.			
/W(100)		pos.			
/W(111)		pos.			
J/W (110)		zero			
/W(100)		neg.			
/W(111)		neg.			
Cl/W(211)	T_R	neg.	FE	1972	5.62
Cl/W(110)					
Ce/W	T_A	neg.	FE	1971	5.138
(Cs+O)/W(112)	T_R	neg.	CPD	1971	5.134
(Cs+O)/W(100)	T_A	neg.	CPD	1973	5.135
Cs/W(100)	T_R	neg.	CPD	1971	5.123
Cs/W(110)	T_R	neg.			

Table 5.3 (continued)

System	Study acc. to: T_R or T_A	Sign (s) of $(\partial\phi/\partial N_a)_{N_a\to0}$	Method of measur.	Year	Reference
Cs/W(111)	T_R	neg.	CPD	1975	5.76
Cs/W(100)	T_R	neg.	CPD	1973	5.81
H/W(100)	T_R	pos.			
Cs/W(100)	T_R	neg.	CPD	1972	5.83
/W(110)	T_R	neg.			
/W(111)	T_R	neg.			
/W(112)	T_R	neg.			
Ba/W(100)	T_R	neg.			
/W(110)	T_R	neg.			
/W(111)	T_R	neg.			
/W(112)	T_R	neg.			
Cs/W(110)	T_R	neg.	FE	1969	5.89
/W(100)	T_R	neg.			
/W(211)	T_R	neg.			
/W(111)	T_R	neg.			
Cs/W(100)	T_R	neg.	CPD	1971	5.30
K/W (100)	T_R	neg.			
Cs/W(100)	T_R	neg.	CPD	1972	5.98
Cs/W(100)	T_R	neg.	CPD	1974	5.105
(O+Cs)/W(100)	T_R	neg.			
CH_4/W(100)	T_R	neg.	CPD	1972	5.63
C_2H_6/W(100)	T_R	neg.			
C_3H_8/W(100)	T_R	neg.			
CH_4/W(100)	T_R	-	CPD	1971	5.152
(H+CO)/W(100)	T_A	pos., neg.	CPD	1971	5.88
CH_4/W(111)		neg.	CPD	1972	5.119
O_2/W(111)		pos.			
H_2/W(111)		pos.			
CH_4/W(110)	T_R	pos.	TE	1972	5.61
CO/W(210)	T_A	pos., neg.	CPD	1972	5.41
H/W(112)	T_A	pos.	CPD	1970	5.37

Table 5.3 (continued)

System	Study acc. to: T_R or T_A	Sign (s) of $(\partial\phi/\partial N_a)_{N_a\to 0}$	Method of measur.	Year	Reference
H/W(001)	T_R	pos.	FE	1973	5.53
/W(011)	T_R	pos.			
/W(121)	T_R	pos.			
H/W(110)	T_A	neg.	CPD	1974	5.101
/W(100)	T_A	pos.			
/W(112)	T_A	pos.			
/W(111)	T_A	neg.			
C_2H_4/W(110)	T_A	neg.			
/W(100)	T_A	neg.			
/W(112)	T_A	neg.			
/W(111)	T_A	neg.			
O/W(110)	T_R	pos.	FE	1972	5.141
/W(112)	T_R	pos.			
/W(111)	T_R	pos.			
/W(013)	T_R	pos.			
O/W	T_A	pos.	CPD	1971	5.146
O/W(100)	T_R	pos.	CPD	1974	5.105
O/W(110)	T_R	-	CPD	1971	5.111
O/W(100)	T_A	pos.	CPD	1976	5.6
O/W(100)	T_A	pos.	CPD	1971	5.94
O/W(110)	T_A	neg.	CPD	1975	5.57
O/W	T_A	pos.	CPD	1976	5.11
O/W	T_A	pos.	CPD	1971	5.47
Ge/W	T_R	pos.	FE	1973	5.95
I/W(100)	T_R	neg.	CPD	1969	5.153
Br/W(100)	T_R	pos.			
Cl/W(100)	T_R	pos.			
(I+Cs)/W(100)	T_R	-			
(Cl+Cs)/W(100)	T_R	-			
N/W(100)	T_A	neg.	CPD	1971	5.29
/W(210)	T_A	pos.			
/W(310)	T_A	pos.			

Table 5.3 (continued)

System	Study acc. to: T_R or T_A	Sign (s) of $(\partial\phi/\partial N_a)_{N_a \to 0}$	Method of measur.	Year	Reference
Na/W(011)	T_A	neg.	CPD	1970	5.91
Na/W(112)	T_R	neg.	CPD	1973	5.73
Na/W(110)	T_R	neg.	FE	1968	5.96
/W(112)	T_R	neg.			
/W(100)	T_R	neg.			
/W(111)	T_R	neg.			
(Ns+0)/W(112)	T_R	neg.	CPD	1971	5.102
I/W(110)	T_A	neg.	CPD	1974	5.104
(Ge,Si)/W(110)	T_A	pos.	FE	1974	5.110
/W(211)	T_A	pos.			
/W(100)	T_A	pos.			
Li/W(112)	T_R	neg.	CPD	1973	5.70
Li/W(110)	T_R	neg.	CPD	1974	5.74
/W(112)	T_R	neg.			
/W(111)	T_R	neg.			
/W(100)	T_R	neg.			
(Mg+0)/W	T_A	pos., neg.	CPD	1973	5.118
(Ca+0)/W	T_A	pos., neg.			
(Sr+0)/W	T_A	pos., neg.			
(Ba+0)/W	T_A	pos., neg.			
K/W(112)	T_R	neg.	CPD	1974	5.75
K/W(110)	T_R	neg.	FE	1975	5.87
/W(112)	T_R	neg.			
/W(100)	T_R	neg.			
/W(111)	T_R	neg.			
Kr/W(110)	T_R	neg.	FE	1971	5.124
/W(100)	T_R	neg.			
/W(111)	T_R	neg.			
/W(112)	T_R	neg.			
Sr/W(110)	T_A	neg.	FE	1974	5.130
Ba/W(110)	T_A	neg.			
Si/W	T_A	neg.	CPD	1973	5.44
Ge/W	T_A	zero			

Table 5.3 (continued)

System	Study acc. to: T_R or T_A	Sign (s) of $(\partial\phi/\partial N_a)_{N_a\to0}$	Method of measur.	Year	Reference
Si/W	T_A	pos.	FE	1971	5.50
Si/W(110)	T_R	pos.	FE	1974	5.69
/W(100)	T_R	pos.			
/W(111)	T_R	pos.			
/W(112)	T_R	pos.			
/W(113)	T_R	pos.			
/W(116)	T_R	pos.			
Sr/W(110)	T_A	neg.	CPD	1974	5.67
Sc/W(110)	T_A	neg.	TE	1971	5.68
/W(100)	T_A	neg.			
/W(111)	T_A	neg.			
S/W	T_A	-	TE	1974	5.143
Se/W	T_A	pos.	FE		
Th/W(100)	T_A	neg.	CPD	1970	5.78
Ti/W	T_A	neg.	FE	1974	5.48
Mo/W(110)	T_A	neg.	CPD	1972	5.45
Pb/W(110)	T_A	neg.	CPD	1975	5.31
/W(100)	T_A	neg., pos.			
Ti/W	T_A	pos.	FE	1971	5.39
U/W(110)	T_A	neg.	FE	1971	5.49
/W(112)	T_A	neg.			
/W(100)	T_A	neg.			
/W(111)	T_A	neg.			
/W(113)	T_A	neg.			
/W(110)	T_A	neg.			
Xe/W(111)	T_R	neg.	CPD	1974	5.56
Yb/W	T_A	neg.	FE	1970	5.122
Nd/W	T_A	neg.			
Zr/W(110)	T_A	neg.	FE	1971	5.125
/W(112)	T_A	neg.			
/W(100)	T_A	neg.			
/W(111)	T_A	neg.			
/W(310)	T_A	neg.			

6. Work Function of Alloys

J. Hölzl

The investigation of binary compounds is at present an area of quite high activity as it is closely connected with an intense interest in application to techniques such as catalysis; in petrochemistry, for example, many compounds are synthesized by using bimetallic catalysts /6.1,2/.

One very severe problem, however, inherent to an alloy study of this kind, has to be solved for each case separately, and that is the determination of the actual surface concentration of the alloy constituents. Various tools such as Auger spectroscopy or XPS have been applied successfully to surface composition determination, and in most of those experiments enrichment phenomena in the surface region have been observed /6.3/. The WF may play an important role in this context since it includes to a large extent surface information. In this sense, e.g., enrichment processes, chemisorption, and diffusion phenomena can be followed easily by studying the WF behavior of the alloy surface in question.

In this chapter we shall elaborate on how far a WF measurement and or a variation in WF as a function of certain alloy parameters can contribute to these studies. Before discussing these experiments (Sects. 6.3,4), a brief review of the theory (Sect. 6.1), and some basic alloy preparation procedures (Sect. 6.2), will be sketched.

6.1 Summary of Theoretical Treatments of Alloy Systems

The old Rigid Band Model (RBM) of MOTT /6.4/ has been used for many years to explain the electronic structure of alloys; however, in many cases this model fails and can therefore not be regarded as an adequate description of such systems /6.5,6/. It predicts common d- and sp-bands of the alloy constituents leading to a single "rigid" form of density of states N(E) for all metals in the same period of the periodic table. One of the main features is that on alloying N(E) is filled up to a certain level according to the number of electrons representing the fraction of each constituent, with the result that the individual character of the N(E) distribution of each alloy component is cancelled out. In contrast it has been found that, especially for systems like CuNi and AgPd, the more recent developed Coherent Potential Approximation (CPA) theory /6.7/ agrees much better with experimental results. In particular this theory makes use of the localized character of electrons associated with the components of a given alloy, compared with the "collectivization" used in the RBM theory.

The most sophisticated theoretical contribution to the problem of the WF of alloys seems to be that of GELATT and EHRENREICH /6.8/ on the AgAu system. A detailed description of their work is given in the theoretical part of this review.

GELATT and EHRENREICH considered a strictly periodic, infinitely extended, lattice containing Au and Ag atoms randomly distributed over the lattice sites, with a given Au atom concentration ($0 \leq x \leq 1$). Attention is focussed on the density of states $g(E,x)$ of the hybridized s-d-band complex. It determines the concentration dependent Fermi level $\varepsilon_F(x)$ via the equation

$$\int_{-\infty}^{\varepsilon_F(x)} g(E,x)dE = xN^{Au} + (1-x)N^{Ag}, \tag{6.1}$$

where N^{Au} and N^{Ag} are the numbers of valence electrons in the pure metals. The pure metal densities of states are adjusted to the respective band structures and the Fermi energy difference $\varepsilon_F^{Ag} - \varepsilon_F^{Au} = \varepsilon_F(0) - \varepsilon_F(1)$ is fitted to the experimentally measured contact potential difference (CPD) of 0.9 eV /6.9/. For $0 \leq x \leq 1$ GELATT and EHRENREICH used the CPA to calculate $g(E,x)$ and thereby $\varepsilon_F(x)$ according to (6.1). If the d-band shift (see theoretical section) is included, the theoretical curve $-\varepsilon_F(x)$ has the same curvature as the experimentally measured CPD, but the experimentally found deviation from linearity is about three times larger than that calculated, see Sect. 6.3. For an explanation of this result it must be kept in mind that the surface contribution of the WF was not explicitly considered.

The reader interested in further details is referred to the original literature /6.7,8/.

6.2 Preparational Procedure and Usefulness of Concentration Graphs

When using alloy bulk single crystals or polycrystalline foils the main emphasis has to be laid upon surface cleaning procedures that avoid the formation of artificial surface concentration gradients, e.g., due to selective sputtering. In the same way the application of elevated temperatures may lead to surface enrichment, especially when dealing with alloys whose constituents differ in vapor pressure.

When working with alloy thin films, many kinds of preparation are available in principle which have been extensively treated by MAISSEL and GLANG /6.10/. Concerning surface cleaning and annealing one has, of course, to contend with the same problems as in the bulk case mentioned above. The most important alloy preparation procedures are listed briefly in the following:

1) The sample is prepared outside the device used to study the WF. This is done mainly in the case of refractory alloy systems /6.11/ where higher temperature effects are to be studied. Thereby special care must be taken to control the surface conditions with respect both to the absence of contamination and a possible concentration gradient normal to the surface of the sample.

128

2) The alloy specimen is prepared _in_ the same vacuum system as is used to study
the WF (in situ preparation) /6.12/. In this case the alloy components are
often deposited simultaneously onto a chemically neutral substrate (e.g.,
Pyrex glass) under UHV conditions. A subsequent heating process for some hours
at a chosen temperature is usually sufficient to form the alloy. This method
which has been used frequently in recent investigations, has been developed
to a high degree of reproducibility. Removal of contaminants from the alloy
surface is obviated and at the same time a quantitative control of bulk and
surface composition is possible.

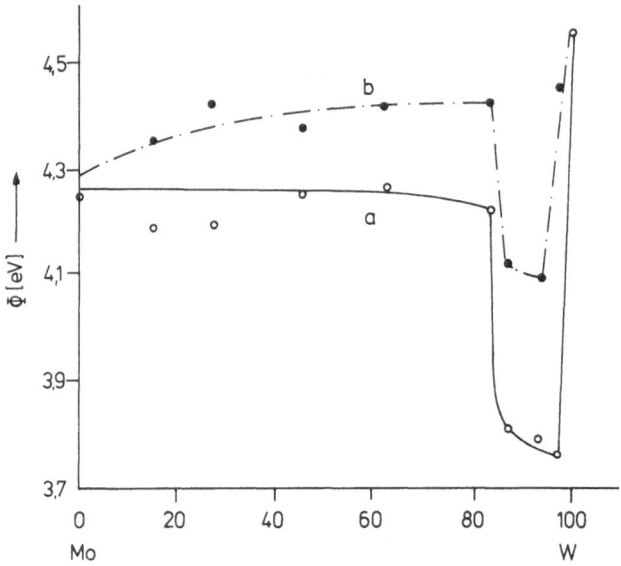

Fig. 6.1. Dependence of the WF of the Mo/W alloy at 1500 K on the tungsten content;
curve a: after heating at 2000 K for 50 hrs; curve b: after heating at 2300 K for
15 hrs. From B.CH. DYUBUA et al. /6.11/

If the WF is plotted as a function either of bulk or of surface composition of
one component (in mole fraction or percent by weight) the resultant diagrams are
called concentration graphs. Two typical concentration graphs are presented for
the two different methods of preparation mentioned above. Firstly, DYUBUA et al.
/6.11/ have made a thermionic study of the solid solution MoW. The alloy discs were
prepared outside the measuring system according to method (1) as described above.
In Fig. 6.1 curves (a) and (b) represent the Φ-dependence of bulk composition in
atomic % after heating the sample to 2000 K for 50 hours and 2300 K for 15 hours
respectively.

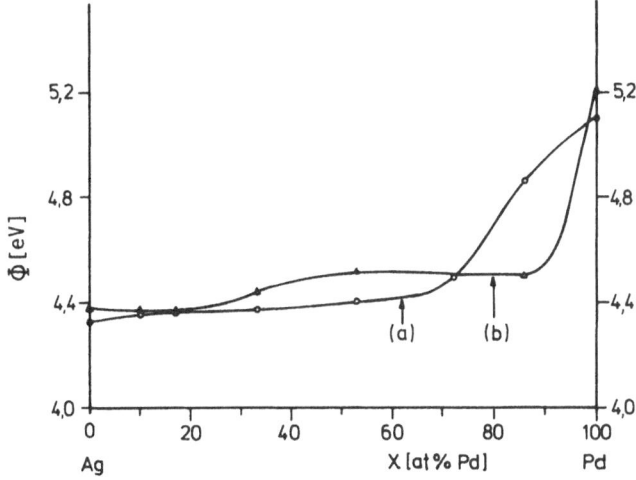

Fig. 6.2. The WF of AgPd alloy film, as a function of the overall composition, curve a: freshly evaporated film; curve b: equilibrated at 573 K. After R. BOUWMAN et al. /6.13/

Secondly, in Fig. 6.2 a result using an in situ experiment is demonstrated. There, BOUWMAN et al. /6.13/ investigated the AgPd alloy system. The preparation of the sample was carried out at a pressure of 10^{-10} Torr by simultaneously evaporation of the two metals onto a support (pyrex) cooled to 78 K. After finishing the evaporation the WF vs. overall composition was measured (freshly evaporated film, curve a), following which the film was equilibrated at 573 K for 16 hours (curve b). The bulk composition of the film was determind by weighing carefully the respective filaments before and after evaporation. The WF was measured photoelectrically. From the result of these two experiments it can be stated that

(I)   In general a linear relationship between WF and overall composition of the bulk does not exist.

(II) The concentration graph of an alloy system is altered drastically by varying the preparation conditions of the sample.

This WF behavior vs. overall composition of the sample is to a large extent connected with the different bulk and surface diffusion processes which can give rise to those crystallographic and/or compositional variations that are dependent on preparation and annealing procedures. In particular it is well known that alloys (mainly films) equilibrated at a fixed temperature generally exhibit a bulk compo - sition different from that of the surface and that as a consequence a WF experiment can give additional information with respect to the ratio of bulk to surface composition.

6.3 Work Function and Surface Composition of Alloys

SACHTLER /6.14/ has pointed out that in the absence of chemisorbing gases the differences between the bulk and surfaces composition of an alloy are due to

(I) the lowering of total free energy by enrichment of the surface with that
   component which causes a decrease in surface energy;

(II) the formation of two stable and coexisting phases.

A theoretical treatment of surface enrichment has recently been given by WILLIAMS
and NASON /6.15/. Clearly measurement of the WF of alloy samples cannot yield all
the desired information about their surfaces. The situation is too complicated.
Nevertheless SACHTLER and DORGELO /6.16/ have developed formally the so-called
chemisorptive titration method, where the $\Delta\Phi$ values of an alloy system can be correlated within certain limits with the surface compositions of the constituents.
The method is based on the principle that the WF of a pure metal is in many cases
changed drastically by the adsorption of a gas, the effect varying in magnitude and
sign depending on the nature of the adsorbate used. This so-called titration method
is based on the assumption that (under favorable circumstances) a gas can be found
which is adsorbed by only one of the alloy constituents. If so, the detectable
amount of the gas chemisorbed on the surface under examination, is then assumed to
be proportional to the number of atoms of the one constituent. Furthermore if $\Delta\Phi$,
the variation of WF, is taken to be proportional to the number of adsorbed gas particles, then the actual surface fraction of this alloy component, which may differ
from that of the bulk, can be measured. Several workers have used this method in
the past and have obtained some valuable results /6.17/.

According to BOUWMAN et al. /6.13/, there are in principle three possible idealized curves (I, II and III) which can be obtained from such an experiment (see
Fig. 6.3), based on their observations on the AgPd system. A straight line (curve I)
would be expected if the surface composition is equal to that of the bulk and if
there is no mutual depolarisation between the adsorbed gas dipoles. A curve showing
a convex behavior with respect to the straight line (curve II) should be obtained
if the surface is enriched with palladium atoms (again depolarisation effects are
neglected). Finally a concave behavior of the $\Delta\Phi$ dependence (curve III) should
indicate that the surface is enriched with silver.

At first sight the titration method seems to be highly attractive as it provides
pure surface information. A detailed analysis of various adsorption systems, however, shows that the assumptions of this model are frequently incorrect /6.18/ for
reasons which have been outlined recently by SACHTLER /6.14/, viz.

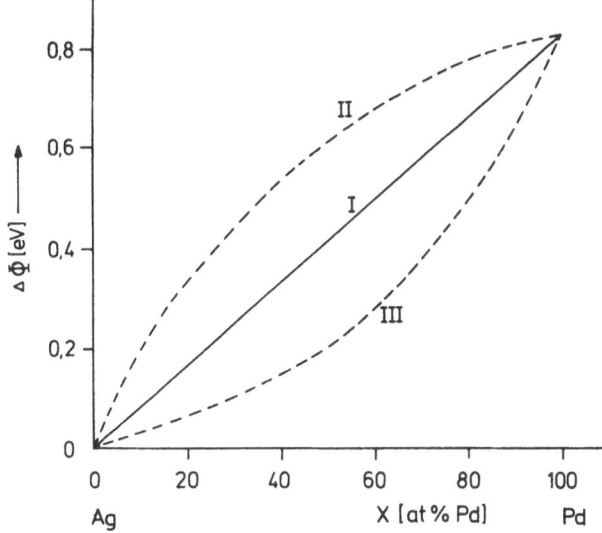

Fig. 6.3. Possible shapes of the Δφ - vs. X_{AgPd} curve. (Explanation in the text). After R. BOUWMAN et al. /6.13/

(I)   The chemisorptive character of a metal atom is more or less affected by its neighbors ("ligand" effect).

(II)   In most cases the chemisorption bond is not localized at one surface atom only, so that a variety of ad-complexes may result and no unique coordination of the number of adsorbate gas-particles to adsorption sites is possible ("ensemble" effect).

(III) As a result of the presence of an adsorbate on the surface, the surface composition of the system may even be changed (de-alloying process; "corrosive" chemisorption), depending on the nature of the metal-adatom bond which may differ strongly for the alloy constituents.

With these problems in measuring surface composition in mind, new techniques have been developed. For example, BRONGERSMA and BUCK /6.19/ have used low energy ion scattering, and numerous authors have applied suitable modification of the AES technique /6.18,20-24/. Recently TAKASU et al. /6.21/ studied CuNi alloys after annealing at 800°C for 8 hours. After the removal of about one hundred surface layers by means of ion bombardment a clean surface was obtained on the sample (p=10^{-10} Torr). Using AES, residual surface impurities as well as the surface composition of the alloy were determined. The WF was then measured photoelectrically

132

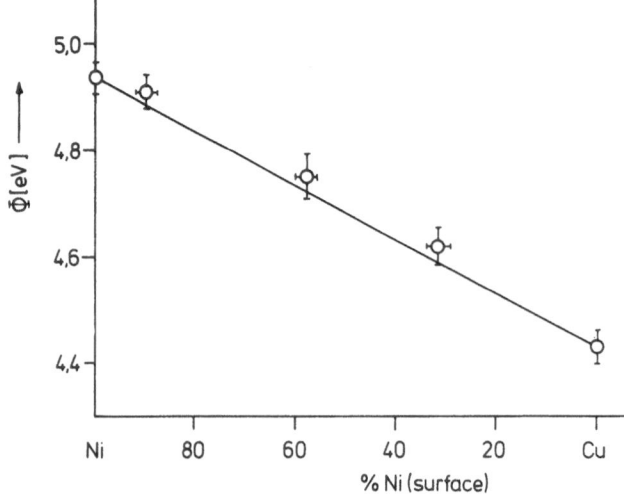

Fig. 6.4. WF of CuNi alloys cleaned by argon ion bombardment vs. their surface composition. After Y. TAKASU et al. /6.21/

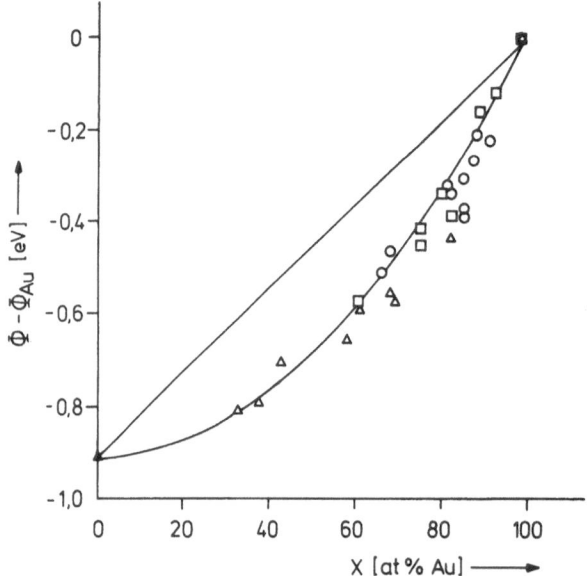

Fig. 6.5. WF difference between alloys AgAu and a gold reference ($\Phi-\Phi_{Au}$) as a function of surface composition. Films were evaporated on a glass substrate. Three different samples (o;Δ;□) were used. After S.C. FAIN Jr. et al. /6.23/

and correlated with the surface composition. The result is shown in Fig. 6.4.
Evidently in this system ϕ is linearly dependent on the alloy composition of the
surface in contradiction to other investigations /6.25/. In closing Sect. 6.3 a
comparison of results obtained theoretically by GELATT and EHRENREICH /6.8/, and
those obtained experimentally by FAIN and McDAVID /6.23/ should be discussed. GELATT
and EHRENREICH have calculated the variation in the bulk contribution $\bar{\mu}$ of the WF
of AgAu alloys as a function of composition. The surface contribution $\Delta\phi$ however
is assumed to be the same for the two constituents Ag and Au, (see Sect. 6.2). In
the experiments of FAIN et al. /6.23/ AgAu alloy films were prepared very carefully.
By means of electron escape technique the films were shown to have the same bulk
and surface composition. The latter was determined by comparing AES signals from
the alloys with signals from pure metal reference standards. The variation of WF
for various AgAu films evaporated on glass substrates as a function of surface
composition is shown in Fig. 6.5. The variation in WF of these alloys falls significantly below the linear interpolation between the two pure metals and is described
approximately by $\phi = \phi_{Ag} + 0.91x^2$ where x is the fractional gold content of the surface. How much of the nonlinearity is due to the surface contribution $\Delta\phi$ of the
alloy WF cannot be concluded from this experiment. Recently FAIN et al. /6.24/ have
studied the system CuAu in a similar experiment.

DAVIDSON and FAIN /6.26/ have calculated the WF of both AgAu and CuAu clusters by
means of an extended Hückel molecular orbital theory. In certain cases the calculated
ionisation potentials show variation with alloy composition similar to that of the
WF of bulk samples.

6.4 Work Function and Other Alloy Characteristics

6.4.1 Work Function and Bulk/Surface Properties

A great deal of work has been devoted to the correlation of WF data with alloy
phase diagram features.

MALOV et al. /6.27/ have studied both very simple systems, as for example a solid
solution of alkali metals like CsRb, and systems containing intermetallic compounds
(e.g., NaCs, etc.). In Fig. 6.6 both the phase diagram and the WF vs. bulk composition for the system RbCs are indicated. All the measurements have been carried out
photoelectrically at a temperature of T = 25°C. In Fig. 6.7 additional results are
presented for the system NaCs. Again both the phase diagram and the WF isotherm at
T = -90°C are indicated. Neglecting concentrational deviations in the outermost
alloy layers, which in fact means that the bulk phase diagram is supposed to remain
unchanged in the surface region, the authors deduce two main results from their
investigations, viz.

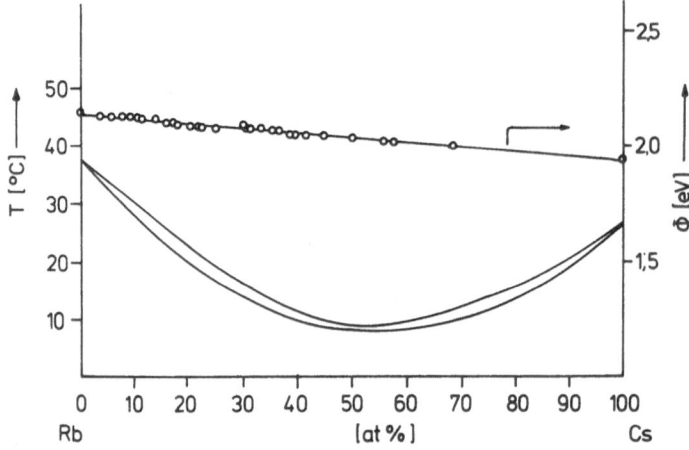

Fig. 6.6. Phase diagram for the system rubidium/cesium and the WF relationship of that system. /6.27/

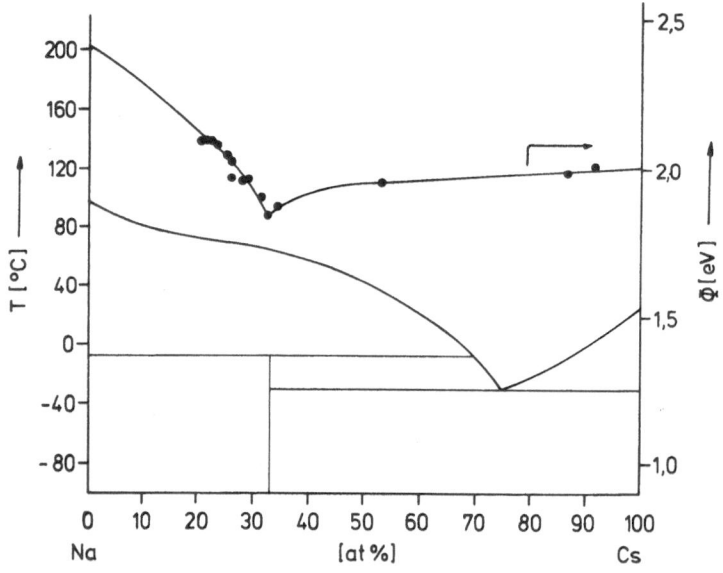

Fig. 6.7. Phase diagram and WF vs. concentration for the system NaCs at -90°C. /6.27/

(I)   Regarding the solid solution RbCs no noticeable deviation from linearity
      in the transition region from solid to liquid solution is observed.

(II) In the case of NaCs the isotherm for the WF at -90°C reveals a pronounced
      minimum in the region of Na_2Cs compound composition where the Φ value
      (=1.85eV) is considerably smaller than that for pure Cs(=1.95eV).

Up to now we have reported the close correlation of some WF characteristics with
features of (mostly polycrystalline) bulk propertiers (phase diagram). It now seems
to be necessary to point out that a combined LEED AES WF experiment can contribute
to an understanding of surface ordering characteristics of monocrystalline alloy
samples, characteristics which may differ from those of the bulk. POTTER and BLAKELY
/6.28/ have investigated a CuAu alloy single crystal using just such an experimental
combination. They correlated with bulk composition the relative Auger signal heights,
the occurence of surface ordering, and the variation of WF with crystallographic
orientation. One of the prominent results of this study, shown in Fig. 6.8, is the
dependence of contact potential difference $\Delta\Phi = \Phi_{crystal.} - \Phi_{refer.}$ on bulk composition and on crystal plane orientation. Some general conclusions can be drawn from
the experiments of POTTER and BLAKELY. In agreement with the surface roughness model
as discussed in Sect. 2.2 of this article the general trend, with the exception of
gold, is found to be

$$\Phi(111) > \Phi(100) > \Phi(110)$$

This orientation dependence is qualitatively consistent with the expected variation
of the surface dipole moment /6.29,30/.

Both the average WF Φ of the alloy and the WF of a given plane Φ (hkl) increase
as the bulk Au concentration increases. To explain this the authors emphasize two
competing effects. As the average valence electron density decreases in proceeding
from Cu to Au, the (jellium) surface electronic dipole is expected to decrease.
Simultaneously the "free" electron Fermi energy should also decrease implying an
increase in the bulk contribution to WF.

An additional observation made by these authors is indicated in Fig. 6.8. The
dashed line represents the decomposition potential for electrolytic dissolution
of polycrystalline CuAu alloys in a 5 N HCl as determined by MÜLLER et al. /6.31/.

In the last part of this section we shall deal with a unique method of measuring
diffusion characteristics of an alloy film. It will be shown in detail that with
the use of a WF experiment both the diffusion coefficient and its temperature dependence can be obtained.

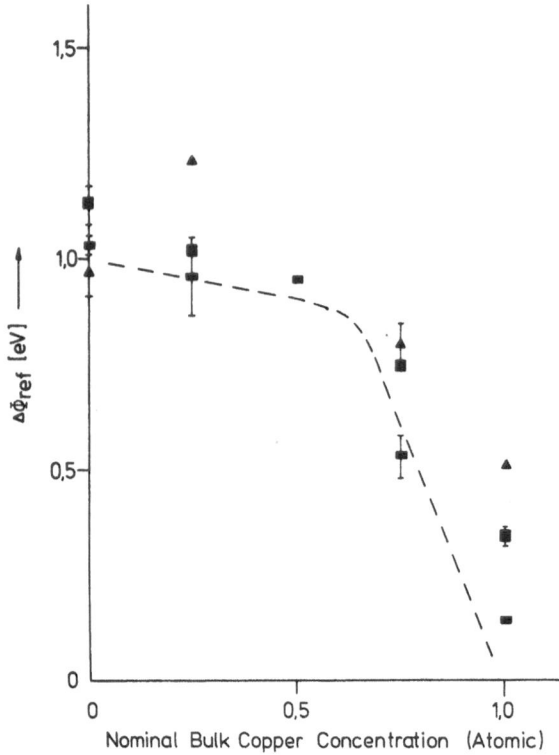

Fig. 6.8. Dependence of contact potential difference on composition and crystal plane for the system CuAu. Δ:(111) surface; □:(100) surface and □□:(110) surface; the dashed line is explained in the text. /6.28/

6.4.2 Use of Work Function Measurement for Obtaining Thin Alloy Film Diffusion Parameters

Very few techniques are available for obtaining reliable experimental results on diffusion data for thin film systems /6.32-35/. THOMAS and HAAS /6.36/ however, have developed an efficient method of measuring not only the diffusion coefficient D but also its temperature dependence. They have demonstrated this technique for the system CrAu. Onto a thin (0.5 x 0.8 x 0.05 cm^3) Ir single crystal in (111) orientation the authors evaporated thin films of either CrAu or AuCr. Most of the diffusion measurements were performed with the system in Fig. 6.9. In it an Aufilm with thickness L (=150 Å) was deposited on top of a Crfilm with thickness H_a (≈ 15 Å). A retarding potential technique /6.37/ was used to measure the WF Φ of

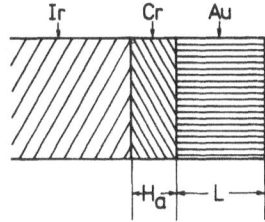

Fig. 6.9. Thin film system CrAu evaporated onto a Ir(111) crystal. From the experiment of THOMAS et al. /6.36/

the sample which could be kept at a fixed temperature. A carefully performed preliminary experiment revealed a linear relationship between Φ and the concentration of Cr atoms C_{Cr}, within a certain range ($0 < C_{Cr} \lesssim 0.32$), thus

$$\Phi = (\Phi_{Cr} - \Phi_{Au}) C_{Cr} + \Phi_{Au}. \qquad (6.2)$$

The time-dependent concentration of Cr atoms $C(t)$ with respect to C_f (the final concentration of Cr atoms at the surface for $t \to \infty$) is therefore given by

$$\frac{C(t)}{C_f} = \frac{\Phi(t) - \Phi_{Au}}{\Phi_f - \Phi_{Au}}, \qquad (6.3)$$

$\Phi(t)$ and Φ_f being the WF's of the surface at time t and at $t \to \infty$, respectively, Φ_{Au} represents the WF of the clean Aufilms. Note that in the course of this experiment the absence of contamination of the surface was monitored very carefully by means of AES.

Taking the Crfilm as sufficiently thin ($H_a \to 0$) Fick's second law $\partial C/\partial t = D \partial^2 C/\partial x^2$ can be applied to this system (x denoting the spatial coordinate in one dimension). The solution for this special problem (finite film thickness) is found by a superposition of solutions for the semi-infinite medium. From that one obtains

$$\frac{C(t)}{C_f} = \left(\frac{L^2}{4Dt}\right)^{1/2} \sum_{n=0}^{\infty} \exp\left(\frac{-(2n+1)^2 L^2}{4Dt}\right) \qquad (6.4)$$

where L is the thickness of the Aufilm, $C(t)$ the surface concentration of Cr atoms at time t and D the diffusion coefficient.

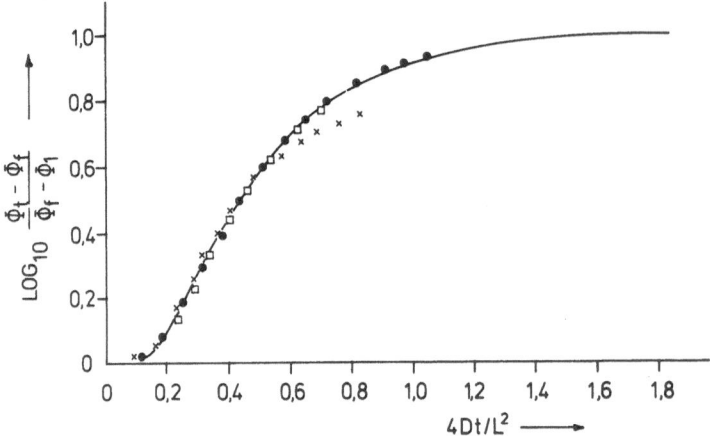

Fig. 6.10. Plot of normalized WF changes vs. $4Dt/L^2$ for diffusion of Cr into Au overlayer films at different Au thicknesses. /6.36/

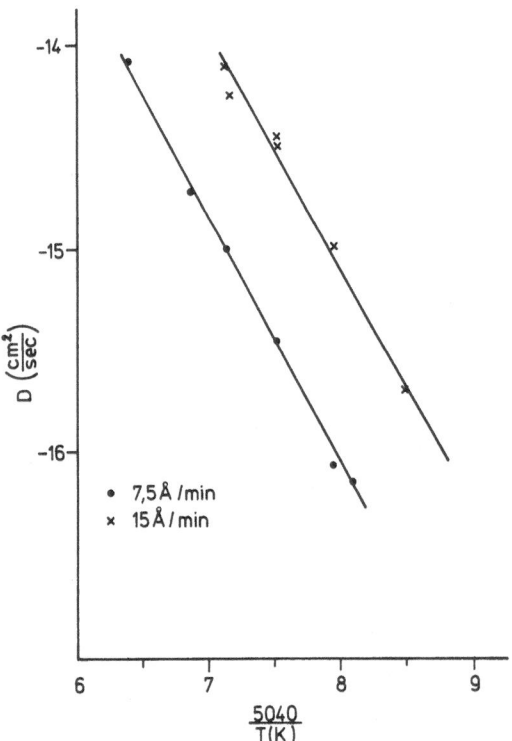

Fig. 6.11. ARRHENIUS plots of D vs. 5040/T for Cr through 150-Å Aufilms at evaporation rates of 7,5 and 15 Å/min. /6.36/

Combining (6.3) and (6.4), the time-dependent quantity $(\Phi(t) - \Phi_{Au}) / (\Phi_f - \Phi_{Au})$ can now be plotted vs. $4Dt/L^2$. This is done in Fig. 6.10, (solid line). The experimental points for the various thicknesses are fitted to the curve by a proper choice of D. Finally in Fig. 6.11 a few results obtained with this method are presented.

Acknowledgement

The authors are greatly indebted to Prof. J.C. Rivière for his invaluable help in critically reading the manuscript at various stages of completion. He never ceased to patiently draw the authors' attention to various insufficiencies of the manuscript. In addition he has taken the trouble of a careful linguistic revision of the text. It is a pleasure to acknowledge the interest which Prof. L. Fritsche has taken in discussing numerous details of this article. He has always considered it challenging and rewarding to find out interconnections between various aspects of theoretical and experimental work. One of the authors (F.K.S.) thanks Prof. H. Bross for continuous support. The other author (J.H.) is very much obliged to P. Schrammen for his substantial help in the formulation of the chapter on Experimental Procedure. The sections dealing with chemisorption and alloys have greatly benefited from a number of suggestions made by Dr. K. Christmann. We are grateful to Prof. E.G. Bauer for his experienced advice on various open questions. The manuscript has also been influenced by valuable suggestions due to Dr. B.E. Nieuwenhuys and Dr. H. Wagner. The preparation of this article would not have been possible without the help of Dr. G. Meister and of all the co-workers of one of the authors (J.H.). The authors are deeply grateful to Frau G. Humburg for her technical services which she performed with infinite skill during the long time this article was in the process of completion.

References

1.1 C. Herring, M.H. Nichols: Rev. Mod. Phys. 21, 185 (1949)
1.2 V.S. Fomenko: Handbook of Thermionic Properties, ed. by G.V. Samsonov (Plenum Press, Data Devision, New York 1966)
1.3 V.S. Fomenko: Emission Properties of Materials, 3rd ed. (Naukova Dumka, Kiev 1970) (in Russian)
1.4 J.C. Rivière: "Work Function: Measurements and Results" in Solid State Surface Science, Vol. 1, ed. by M. Green (Decker, New York 1969), Chap. 4, pp. 179-289
1.5 H.B. Michaelson: J. Appl. Phys. 48, 4729 (1977)
1.6 G.A. Haas, R.E. Thomas: "Thermionic Emission and Work Function", in Techniques of Metals Research, Vol. 1, ed. by E. Pasaglia (Wiley-Interscience, Chichester, Sussex 1972), pp. 91-262
1.7 N.D. Lang: "The Density-Functional Formalism and the Electronic Structure of Metal Surfaces" in Solid State Physics, Vol. 28, ed. by H. Ehrenreich, F. Seitz, D. Turnbull (Academic Press, New York, London 1973)
2.1 W. Gordy, W.J.O. Thomas: J. Chem. Phys. 24, 439 (1956)
2.2 D. Steiner, E.P. Gyftopoulos: "An Equation for the Prediction of Bare Work Functions"; in Proc. 27th Conf. on Phys. Electronics, Cambridge, Mass. 1967, pp. 160-168
2.3 E. Wigner, J. Bardeen: Phys. Rev. 48, 84 (1935)
2.4 J. Bardeen: Phys. Rev. 49, 653 (1936)
2.5 P. Hohenberg, W. Kohn: Phys. Rev. 136, B 864 (1964)
2.6 W. Kohn, L.J. Sham: Phys. Rev. 140, A 1133 (1965)
2.7 L.J. Sham, W. Kohn: Phys. Rev. 145, 561 (1966)
2.8 V. Heine, C.H. Hodges: J. Phys. C: Solid St. Phys. 5, 225 (1972)
2.9 C.H. Hodges: J. Phys. F: Metal Phys. 4, 1961 (1974)
2.10 R.M. Nieminen, C.H. Hodges: J. Phys. F: Metal Phys. 6, 573 (1976)
2.11 N.D. Lang, W. Kohn: Phys. Rev. B 3, 1215 (1971)
2.12 N.D. Lang: "The Density-Functional Formalism and the Electronic Structure of Metal Surfaces"; in Solid State Physics, Vol. 28, ed. by H. Ehrenreich, F. Seitz, D. Turnbull (Academic Press, New York, London 1973) pp. 225-300
2.13 R. Smoluchowski: Phys. Rev. 60, 661 (1941)
2.14 I. Langmuir: J. Am. Chem. Soc. 54, 2798 (1932), see also:
J.B. Taylor, I. Langmuir: Phys. Rev. 44, 423 (1933)
2.15 R.W. Gurney: Phys. Rev. 47, 479 (1935)
2.16 D.M. Newns: Phys. Rev. 178, 1123 (1969),
P.W. Anderson: Phys. Rev. 124, 41 (1961); see also:
S.K. Lyo, R. Gomer: "Theory of Chemisorption"; in Topics in Applied Physics, Vol. 4, ed. by R. Gomer (Springer Berlin, Heidelberg, New York 1975) Chap. 2, pp. 41-62
2.17 J.L. Moran-Lopez, A. Ten Bosch: Surface Sci. 68, 377 (1977)
2.18 J.L. Moran-Lopez, A. Ten Bosch: to be published
2.19 H.E. Albrecht: Phys. Stat. Sol. (a) 9, 125 (1972)
2.20 E.P. Gyftopoulos, J.D. Levine: J. Appl. Phys. 33, 67 (1962)
2.21 E.P. Gyftopoulos, D. Steiner: Report of the 27th Annual Conference on Physical Electronics; Cambridge, Mass. (1972) pp. 169-187
2.22 N.S. Rasor, C. Warner: J. Appl. Phys. 35, 2589 (1964)
2.23 C.D. Gelatt, Jr., H. Ehrenreich: Phys. Rev. B 10, 398 (1974)
2.24 R.G. Forbes: "Re-Exploration of the Concepts of Work Function and Ionization Energy"; in Proc. 7th Intern. Vac. Congr. & 3rd Intern. Conf. Solid Surfaces, Vienna 1977, pp. 433-436
2.25 J.P. Eckmann, L.E. Thomas: Comm. Math. Phys. 41, 175 (1975)
2.26 See, e.g., D. Pines: Elementary Excitations in Solids, (Benjamin, New York, Amsterdam 1963)
2.27 L. Hedin, St. Lundqvist: Effects of Electron-Electron and Electron-Phonon Interactions on the One-Electron States of Solids; in Solid State Physics, Vol. 23, ed. by H. Ehrenreich, F. Seitz, D. Turnbull (Academic Press, New York, London 1969) pp. 1-181

2.28 O. Gunnarsson, B.I. Lundqvist: Phys. Rev. B 13, 4274 (1976)
2.29 E.P. Wigner: Phys. Rev. 46, 1002 (1934). For a correction of the formula
      proposed by Wigner see Ref. 2.26, Eq. (3.58)
2.30 O. Gunnarsson, B.I. Lundqvist, J.W. Wilkins: Phys. Rev. B 10, 1391 (1974)
2.31 A.S. Kompaneets, E.S. Pavlovskii: Zh. Eksp. Teor. Fiz. 31, 427 (1956)
      (English transl.: Sov. Phys.-JETP 4, 328 (1957); see also /2.5/
2.32 M. Rasolt, D.J.W. Geldart: Phys. Rev. Lett. 35, 1234 (1975) and Phys. Rev.
      B 13, 1477 (1976)
2.33 For a review see J.C. Slater: The Self-Consistent Field for Molecules and
      Solids (McGraw-Hill, New York 1974)
2.34 L.J. Sham: Phys. Rev. A 1, 969 (1970)
2.35 F.K. Schulte: Z. Physik B 27, 303 (1977)
2.36 N.D. Lang: Phys. Rev. B 4, 4234 (1971)
2.37 C.H. Hodges, M.J. Stott: Phil. Mag. 26, 375 (1972)
2.38 P.H. Handel: Z. Physik 252, 7 (1972); Phys. Rev. Lett. 28. 596 (1972)
2.39 D.G. Costello, D.E. Groce, D.F. Herring, J.WM. McGowan: Phys. Rev. B 5, 1433
      (1972)
2.40 B.Y. Tong: Phys. Rev. B 5, 1436 (1972)
2.41 C.H. Hodges, M.J. Stott: Phys. Rev. B 7, 73 (1973)
2.42 C.H. Hodges, M.J. Stott: Solid State Comm. 12, 1153 (1973)
2.43 R.M. Nieminen, C.H. Hodges: Solid State Comm. 18, 1115 (1976)
2.44 F. Seitz: The Modern Theory of Solids (McGraw-Hill, New York 1940) pp. 395-406
2.45 C. Herring, M.H. Nichols: Rev. Mod. Phys. 21, 185 (1949)
2.46 N.D. Lang, W. Kohn: Phys. Rev. B 8, 6010 (1973)
2.47 T. Schneider: Phys. Stat. Sol. 32, 323 (1969)
2.48 G. Allan, M. Lannoo: 2nd Colloque Internationale de Physique et Chemie des
      Surfaces (Brest, France 1970)
2.49 G. Allan "Electronic Structure of Transition Metal Surfaces"; in Electronic
      Structure and Reactivity of Metal Surfaces ed. by E.G. Derouane and A.A. Lucas
      (Plenum Press, New York, London 1976) pp. 45-79
2.50 A. Sommerfeld: Naturwissenschaften 15, 825 (1927); 16, 374 (1928)
2.51 C.F. Von Weizsäcker: Z. Physik 96, 431 (1935)
2.52 H.J. Juretschke: Phys. Rev. 92, 1140 (1953)
2.53 T.L. Loucks, P.H. Cutler: J. Phys. Chem. Solids 25, 105 (1964)
2.54 R.W. Davies: Surf. Sci. 11, 419 (1968)
2.55 J.W. Gadzuk: Surf. Sci. 11, 465 (1968)
2.56 A.J. Bennet, G.B. Duke: "Self-Consistent Charge Densities, Potentials, and
      Work Functions at Metallic Interfaces"; in The Structure and Chemistry of
      Solid Surfaces, ed. by G.A. Somorjai (Wiley, New York 1969) p. 25-1
2.57 A.J. Bennet, C.B. Duke: Phys. Rev. 188, 1060 (1969)
2.58 See, e.g., P. Gombás: Die statistische Theorie des Atoms und ihre Anwendungen
      (Springer, Wien 1949) p. 329
2.59 M. Dubejko, S. Olszowski: Phys. Stat. Sol. 16, 399 (1966)
2.60 J.R. Smith: Phys. Rev. 181, 522 (1969)
2.61 M.M. Pant, M.P. Das: J. Phys. F.: Metal Phys. 5, 1301 (1975)
2.62 C. Warner: IEEE Conf. Proc. TCSC, p. 170 (1971)
2.63 J.R. Smith: "Theory of Electronic Properties of Surfaces"; in Topics in
      Applied Physics, Vol. 4, ed. by R. Gomer (Springer Berlin, Heidelberg, New
      York 1975) Chap. 1, pp. 1-39
2.64 N.D. Lang: Solid State Comm. 7, 1047 (1969)
2.65 N.W. Ashcroft, D.C. Langreth: Phys. Rev. 155, 682 (1967); 159, 500 (1967);
      N.W. Ashcroft: J. Phys. C: Solid State Phys. 1, 232 (1968)
2.66 H.F. Budd, J. Vannimenus: Phys. Rev. Lett. 31, 1218 (1973); 31, 1430 (1973)
2.67 H.F. Budd, J. Vannimenus: Phys. Rev. B 12, 509 (1975)
2.68 H.F. Budd, J. Vannimenus: Phys. Rev. B 14, 854 (1976)
2.69 G.D. Mahan, W.L. Schaich: Phys. Rev. B 10, 2647 (1974)
2.70 N.D. Lang, W. Kohn: Phys. Rev. B 7, 3541 (1973)
2.71 V. Peuckert: J. Phys. C: Solid State Phys. 7, 2221 (1974)
2.72 V. Sahni, J.B. Krieger, J. Gruenebaum: Phys. Rev. B 15, 1941 (1977)
2.73 K.H. Lau, W. Kohn: J. Phys. Chem. Sol. 37, 99 (1976)
2.74 J.H. Rose Jr., H.B. Shore, D.J.W. Geldart, M. Rasolt: Solid State Comm. 19,
      619 (1976) and /2.75/

2.75 J.P. Perdew, D.C. Langreth, V. Sahni: Phys. Rev. Lett. 38, 1030 (1977)
2.76 R. Monnier, J.P. Perdew, Phys. Rev. B 17, 2595 (1978)
2.77 R. Monnier, J.P. Perdew, D.C. Langreth, J.W. Wilkins: Phys. Rev. B 18, 656 (1978)
2.78 J.K. Grepstad, P.O. Gartland, B.J. Slagsvold: Surface Sci. 57, 348 (1976)
2.79 F. Jona: Surf. Sci. 68, 204 (1977)
2.80 G.P. Alldredge, L. Kleinman: J. Phys. F: Metal Phys. 4, L207 (1974)
2.81 G. Paasch, M. Hietschold: Phys. Stat. Sol. (b) 67, 743 (1975)
2.82 M. Hietschold, G. Paasch, P. Ziesche: Phys. Stat. Sol. (b) 70, 653 (1975)
2.83 C. Tejedor, F. Flores: J. Phys. F: Metal Phys. 6, 1647 (1976)
2.84 F.K. Schulte: Surf. Sci. 55, 427 (1976)
2.85 R.C. Jaklevic, J. Lambe, M. Mikkor, W.C. Vassell: Phys. Rev. Lett. 26, 88 (1971)
2.86 D. Stark, P. Zwicknagl: Appl. Phys. 10, 265 (1976)
2.87 C.H. Kelly: M.S. Thesis, Polytechnic Institute of Brooklyn, New York 1954 (unpublished)
2.88 M. Kaplit: IEEE Conf. Proc. TCSC, p. 387 (1976)
2.89 J.R. Smith: Phys. Rev. Lett. 25, 1023 (1970)
2.90 J.G. Gay, J.R. Smith, F.J. Arlinghaus: Phys. Rev. Lett. 38, 561 (1977)
2.91 G.P. Alldredge, L. Kleinman: Phys. Rev. B 10, 559 (1974)
2.92 J.A. Appelbaum, D.R. Hamann: Phys. Rev. B 6, 2166 (1972)
2.93 J.R. Chelikowsky, M. Schlüter, S.G. Louie, M.L. Cohen: Solid State Comm. 17, 1103 (1975)
2.94 S.G. Louie, K.M. Ho, J.R. Chelikowsky, M.L. Cohen: Phys. Rev. Lett. 37, 1289 (1976): Phys. Rev. B 15, 5627 (1977)
2.95 J.A. Appelbaum, D.R. Hamann: Rev. Mod. Phys. 48, 479 (1976)
2.96 H.B. Michaelson: J. Appl. Phys. 48, 4729 (1977)
2.97 J.C. Slater: Quantum Theory of Molecules and Solids, Vol. 2 (McGraw-Hill, New York, St. Louis, San Francisco, Toronto, London, Sydney 1965)
2.98 L. Pauling: The Nature of the Chemical Bond, 3rd ed. (Cornell University Press, Ithaca, New York 1960)
2.99 J. Topping: Proc. Roy. Soc. (London) A114, 67 (1927); see also /2.117/
2.100 I. Higuchi, T. Ree, H. Eyring: J. Am. Chem. Soc. 77, 4969 (1955)
2.101 L.D. Schmidt, R. Gomer: J. Chem. Phys. 45, 1605 (1966)
2.102 J.W. Gadzuk: "Theory of Metallic Adsorption on Real Metal Surfaces"; in The Structure and Chemistry of Solid Surfaces, ed. by G.A. Somorjai (Wiley, New York 1969) p. 43-1; see also references cited in this paper
2.103 J.K. Hartman: "An Approach to Chemisorption within the Anderson Model". Thesis, Cornell University (1970)
2.104 J.P. Muscat, D.M. Newns: Solid State Comm. 11, 737 (1972)
2.105 J.P. Muscat, D.M. Newns: J. Phys. C: Solid State Phys. 7, 2630 (1974)
2.106 Cf. also C.E. Carrol, J.W. May: Surf. Sci. 29, 60 (1972)
2.107 J. Langer, S.J. Vosko: J. Phys. Chem. Solids 12, 196 (1960)
2.108 N.D. Lang, W. Kohn: Phys. Rev. B 7, 3541 (1973)
2.109 N.D. Lang: "Density-Functional Approach to the Electronic Structure of Metal Surfaces and Metal-Adatom Systems"; in Electronic Structure and Reactivity of Metal Surfaces, ed. by E.G. Derouane and A.A. Lucas (Plenum Press, New York, London 1976) pp. 81-111
2.110 H.F. Budd, J. Vannimenus: Phys. Rev. B 12, 509 (1975)
2.111 Cf. J.A. Appelbaum, D.R. Hamann: Phys. Rev. B 6, 1122 (1972)
2.112 A.J. Bennet, L.M. Falicov: Phys. Rev. 151, 512 (1966)
2.113 A.J. Bennet: J. Chem. Phys. 49, 1340 (1968)
2.114 J.W. Gadzuk, J.K. Hartman, T.N. Rhodin: Phys. Rev. B 4, 241 (1971)
2.115 Cf., e.g., W. Brenig: "Chemisorption of H, O and CO on Transition Metals"; in Festkörperprobleme XVII/ Advances in Solid State Physics, ed. by J. Treusch (Vieweg, Braunschweig 1977) pp. 301-317
2.116 Cf., however, E.E. Mola: Surf. Sci. 51, 290 (1975) (Chemisorption of N_2 on Fe) and J. Muller: Izv. Akad. Nauk SSSR Ser. Fiz. 38, 345 (1974) (W.F. change due to physisorption)
2.117 A.C. Hewson, D.M. Newns: Japan. J. Appl. Phys. Suppl. 2, Pt. 2, 121 (1974)

2.118 Cf., e.g., T.B. Grimley: Proc. Phys. Soc. (London) 90, 751 (1967); T.L. Einstein, J.R. Schrieffer: Phys. Rev. B 7, 3629 (1973); N.R. Burke: Surf. Sci. 58, 349 (1976)

2.119 J.R. Smith, S.C. Ying, W. Kohn: Phys. Rev. Lett. 30, 610 (1973); S.C. Ying, J.R. Smith, W. Kohn: Phys. Rev. B 11, 1483 (1975)

2.120 L.M. Kahn, S.C. Ying: Solid State Comm. 16, 799 (1975)

2.121 L.M. Kahn, M. Rasolt: Solid State Comm. 20, 1073 (1976)

2.122 N.D. Lang, A.R. Williams: Phys. Rev. Lett. 34, 531 (1975)

2.123 N.D. Lang, A.R. Williams: Phys. Rev. Lett. 37, 212 (1976)

2.124 K.Y. Yu, J.N. Miller, P. Chye, W.E. Spicer, N.D. Lang, A.R. Williams: Phys. Rev. B 14, 1446 (1976)

2.125 N.D. Lang: Phys. Rev. B 4, 4234 (1971)

2.126 H.B. Huntington, L.A. Turk, W.W. White, III: Surf. Sci. 48, 187 (1975)

2.127 For a description of the coherent potential approximation see H. Ehrenreich, L. Schwartz: "The Electronic Structure of Alloys"; in Solid State Physics, Vol. 31, ed. by H. Ehrenreich, F. Seitz, D. Turnbull (Academic Press, New York, London 1976) pp. 149-286

2.128 S.C. Fain Jr., J.M. McDavid: Phys. Rev. B 9, 5099 (1974)

2.129 E.R. Davidson, S.C. Fain Jr.: J. Vac. Sci. Technol. 13, 209 (1976)

2.130 R.S. Mulliken: J. Chim. Phys. 46, 497, 675 (1949)

2.131 M. Wolfsberg, L. Helmholtz: J. Chem. Phys. 20, 837 (1952)

2.132 P. Fulde, A. Luther, R.E. Watson: Phys. Rev. B 8, 440 (1973)

3.1 S. Berge, P.O. Gartland, B.J. Slagsvold: Surf. Sci. 43, 275 (1974)

3.2 C. Herring and M.H. Nichols: Rev. Mod. Phys. 21, 185 (1949)

3.3 R.H. Fowler: Phys. Rev. 38, 45 (1931)

3.4 L.A. Du Bridge: Phys. Rev. 39, 108 (1932)

3.5 E. Guth, C.J. Mullin: Phys. Rev. 59, 867 (1941)

3.6 D.W. Juenker: Phys. Rev. 99, 1155 (1955)

3.7 D.W. Juenker: J. Appl. Phys. 28, 1398 (1955)

3.8 R.H. Fowler, L.W. Nordheim: Proc. Roy. Soc. (London) A 119, 173 (1928)

3.9 L.W. Nordheim: Proc. Roy. Soc. (London) A 121, 626 (1928)

3.10 M.I. Yelinson, F.F. Dobryakova, V.F. Krapivin, Z.A. Malina, A.A. Yasnopol'-skaya: Radio Eng. Electron Phys. (USSR) 6, 1191 (1961), Radiotekhnika i Elektronika 6, 1342 (1961)

3.11 E.W. Müller: Z. Phys. 120, 261 (1943)

3.11a R. Niedermayer, J. Hölzl: Phys. Stat. Sol. 11, 651 (1965)

3.11b A.R. Hutson: Phys. Rev. 98, 889 (1955)

3.12 Z. Sidorski, I. Pelly, R. Gomer: J. Chem. Phys. 50, 2382 (1969)

3.13 R. Maly: Rev. Sci. Instrum. 44, 1097 (1973)

3.14 T.V. Vorburger, D. Penn, E.W. Plummer: Surf. Sci. 48, 4176 (1975)

3.15 C.E. Kuyatt, E.W. Plummer: Rev. Sci. Instr. 43, 108 (1972)

3.16 A.G. Knapp: Surf. Sci. 34, 289 (1973)

3.17 F.H. Hayes, M.P. Hill, M.A. Lecchini, B.A. Pethica: J. Chem. Phys. 42, 2919 (1965)

3.18 J. Pritchard: Trans. Faraday Soc. 51, 437 (1965)

3.19 J.C.P. Mignolet: Rec. Trav. Chim. Pays Bas: 74, 685 (1955)

3.20 R. Nathan, L.J. Hopkins: Journal of Phys. E.: Sci. Instr. 7, 851 (1974)

3.21 P.A. Anderson: Phys. Rev. 59, 1034 (1941)

3.22 H. Shelton: Phys. Rev. 107, 1553 (1957)

3.23 G.A. Haas, R.E. Thomas: J. Appl. Phys. 34, 3457 (1963)

3.24 G.A. Haas, R.E. Thomas: Surf. Sci. 4, 64 (1966)

3.25 A.A. Holscher: Surf. Sci. 4, 89 (1966)

3.26 W. Thomas (later Lord Kelvin): Phil. Mag. 46, 82 (1898)

3.27 W.A. Zisman: Rev. Sci. Instrum. 3, 367 (1932)

3.28 Y. Petit-Clerc, J.D. Carrete: Appl. Phys. Letters 12, 227 (1968)

3.29 R. Butz, H. Wagner: Appl. Phys. 13, 37 (1977)

3.30 K.A. Macfadyen, T.A. Holbeche: J. Sci. Instru. 34, 101 (1957)

3.31 W.R. Harper: Proc. R. Soc. A 205, 83 (1957)

3.32 Y.L. Yousef, A. Miskriki, S. Aziz, H. Mikhail: J. Sci. Instr. 42, 873 (1965)

3.33 H.H. Kolm: Rev. Sci. Instr. 27, 1046 (1956)

3.34 J.C. Mitchinson, R.D. Pringle, W.E.J. Farvis: J. Phys. (London) E 4, 525 (1971)

3.35 J. Hölzl, P. Schrammen: J. Appl. Phys. 3, 353 (1974)

3.36 N.A. Surplice, R.J. D'Arcy: J. Phys. E 3, 477 (1970)
3.37 J.S.W. De Boer, H.J. Krusmeyer, N.C. Burhoven-Jaspers: Rev. Sci. Instr. 44, 1003 (1973)
3.38 R.E. Simon: Phys. Rev. 116, 613 (1959)
3.39 T.A. Delchar, G. Ehrlich: J. Chem. Phys. 42, 2686 (1965)
3.40 Y. Petit-Clerc, J.D. Carette: Rev. Sci. Instr. 39, 933 (1968)
3.41 B.H. Blott, T.J. Lee: J. Phys. E: Sci. Instr. 2, 785 (1969)
3.42 W. Jaschinski: Diplomarbeit Ruhr-Universität Bochum (1974)
3.43 J. Hölzl, P. Schrammen, G. Porsch: to be published
3.44 S.C. Fain, Jr., L.V. Corbin, II, J.M. McDavid: Rev. Sci. Instr. 47, 345 (1976)
3.45 E. Krimmel, G. Möllenstadt, W. Rothemund: Appl. Phys. Letters 5, 209 (1964)
3.46 J.C. Rivière, in: Solid State Surf. Sci., ed. by M. Green, Vol. 1, p. 179 (Dekker, New York 1969)
3.47 G.A. Haas, R.E. Thomas, in: Techniques of Metal Research, ed. by E. Pasgalia, Vol. 4, Part I (Interscience, New York 1972) p. 91
4.1 E.M. Lavitiskii, I.V. Burov, L.N. Litvak: Sov. Phys. Dokl. 19, 676 (1975)
4.2 H.B. Michaelson: J. Appl. Phys. 48, 4729 (1977)
4.3 C.F. Gallo, W.L. Lama: IEEE Trans. IA-10, 496 (1974)
4.4 I.F. Tamm, D. Blochinzev: Z. Physik 77, 774 (1932)
4.5 W.M.H. Sachtler: Z. f. Electrochem. 59, 119 (1955)
4.6 H. Schade: "Work Function and Sublimation Entropies of the Elements"; in Proc. 7th Intern. Vac. Congr. & 3rd Intern. Conf. Solid Surfaces, Vienna 1977, pp. 437-439
4.7 S.N. Zadumkin, I.G. Sheezukhova, B.B. Al'Chagirov: Fiz. Metal. Metalloved 30, 1313 (1970) (Engl. transl.: Phys. Met. and Metallogr. (GB) 30, 195 (1970)
4.8 H.E. Albrecht: Phys. Stat. Sol. (a) 6, 135 (1971)
4.9 A.K. Mindyuk: Fiz.-Khim. Mekh. Mater. 8, 40 (1972)
4.10 B.F. Rzyanin: Teplofiz. Vys. Temp. (USSR) 11, 34 (1973) (Engl. transl.: High Temp. 11, 30 (1973))
4.11 YU.M. Goryachev, I.A. Podchemyaeva: Phys. Stat. Sol. (b) 56, 443 (1973)
4.12 G.V. Samsonov, I.YA. Kondratov: Isvestiya Vuz Fizika 11, 103 (1968)
4.13 W. Gordy, W.J.O. Thomas: J. Chem. Phys. 24, 439 (1956)
4.14 D. Steiner, E.P. Gyftopoulos: "An Equation for the Prediction of Bare Work Functions"; in Proc. 27th Conf. on Phys. Electronics, Cambridge, Mass. 1967, pp. 160-168
4.15 See also: A.R. Miedema, F.R. de Boer, P.F. de Chatel: J. Phys. F: Metal Phys. 3, 1558 (1973) and Ref. 2.37 (applications to alloys)
4.16 L. Pauling: The Nature of the Chemical Bond, 3rd ed. (Cornell University Press, Ithaca, New York 1960) p. 88
4.16a R.S. Mulliken: J. Chem. Phys. 2, 782 (1934) 3. 573 (1935)
4.17 W. Gordy: Phys. Rev. 69, 604 (1946)
4.18 H.O. Pritchard, H.A. Skinner: Chem. Rev. 55, 745 (1955)
4.19 L. Fritsche, J. Noffke: to be published in Sol. Stat. Comm. 1978
4.20 R. Smoluchowski: Phys. Rev. 60, 661 (1941)
4.21 D.A. Libermann, D.T. Cromer, T.T. Waber: Comp. Phys. Comm. 2, 107 (1971)
4.22 B. Krahl-Urban: Thesis Univ. Köln (1976)
4.23 E. Wigner, J. Bardeen: Phys. Rev. 48, 84 (1935);
J. Bardeen: Phys. Rev. 49, 653 (1936); see also
C. Herring, M.H. Nichols: Rev. Mod. Phys. 21, 185 (1949)
4.24 P. Hohenberg, W. Kohn: Phys. Rev. 136, 864 (1964);
W. Kohn, L.J. Sham: Phys. Rev. 140A, 1133 (1965);
L.J. Sham, W. Kohn: Phys. Rev. 145, 561 (1966)
4.25 N.D. Lang: Solid State Phys., Vol. 28, ed. by H. Ehrenreich, F. Seitz, D. Turn-bull (Academic Press, New York, London 1973) p. 225
4.26 G.P. Aldredge, L. Kleinmann: Phys. Rev. B.10, 559 (1974)
4.27 J.A. Appelbaum, D.R. Hamann: Phys. Rev. B.6, 2166 (1972)
4.28 J.R. Chelikowsky, M. Schlüter, S.G. Louie, M.L. Cohen: Solid State Comm. 17, 1103 (1975)
4.29 S.G. Louie, K.M. Ho, J.R. Chelikowsky, M.L. Cohen: Phys. Rev. Letter 37, 1289 (1976)
4.30 C.H. Hodges: J. Phys. C: Solid State Phys. 5, 225 (1972)

4.31 J.G. Gay, J.R. Smith, F.J. Arlinghaus: Phys. Rev. Lett. 38, 561 (1977)
4.32 N.D. Lang, W. Kohn: Phys. Rev. B1, 4555 (1970);
     N.D. Lang, W. Kohn: Phys. Rev. B3, 1215 (1971)
4.33 E.N. Sickafus: Surf. Sci. 19, 181 (1970)
4.34 B. Heimann, J. Hölzl: Z. Naturf. 27a, 408 (1972)
4.35 K. Besocke: Private Communication 1977
4.36 M. Henzler: "Electron Diffraction and Surface Defect Structure", in Electron
     Spectroscopy for Surface Analysis, ed. by H. Ibach, Topics in Current Physics,
     Vol. 4 (Springer, Berlin, Heidelberg, New York 1977) p. 117
4.37 E.W. Plummer, T.N. Rhodin: Appl. Phys. Letters 11, 194 (1967);
     E.W. Plummer, T.N. Rhodin: J. Chem. Phys. 49, 3479 (1968)
4.38 J.K. Grepstad, P.O. Gartland, B.J. Slagsvold: Surf. Sci. 57, 348 (1976)
4.39 K. Christmann, G. Ertl: Surf. Sci. 60, 365 (1976)
4.40 B. Krahl-Urban, E.A. Niekisch, H. Wagner: Surf. Sci. 64, 5 (1977)
4.41 H. Wagner: in Springer Tracts in Modern Physics Vol. 85
4.42 J.G. Endriz, W.E. Spicer: Phys. Rev. Letters 27, 570 (1971)
4.43 S.A. Flodström, J.G. Endriz: Phys. Rev. Letters 31, 893 (1973)
4.44 G.E. Rhead: Surf. Sci. 1977 (in press)
4.45 S. Yamamoto, K. Susa, U. Kawabe: J. Chem. Phys. 60, 4076 (1974)
4.46 F.K. Schulte: Private Commun. (1975)
4.47 C. Herring, M.H. Nichols: Rev. Mod. Phys. 21, 185 (1949)
4.48 P.O. Gartland, S. Berge, B.J. Slagsvold: Physica Norvegica 7, 39 (1973)
4.49 P. Köhler: Z. Angew. Phys. 21, 191 (1966)
4.50 G. Busch, P. Cotti, H.J. Güntherodt, P. Munz, P. Oelhafen: Helvetica Physica
     Acta 46, 31 (1973)
4.51 JU.I. Malov, M.D. Shebzukhov, V.B. Lazarev: Surf. Sci. 44, 21 (1974)
4.52 R.v. Hill, E.K. Stepanakos, R.F. Tinder: J. Appl. Phys.: 42, 4296 (1971)
4.53 M. Schott, A.J. Walton: Phys. Lett. 60A, 53 (1977)
4.54 A.B. Cardwell: Phys. Rev. 76, 125 (1949)
4.55 G. Comsa, A. Gelberg, B. Iosifescu: Phys. Rev. 122, 1091 (1961)
4.56 M.M. Pant, A.K. Rajagopal: Sol. Stat. Comm. 10, 1157 (1972)
4.57 S.V. Wonsowski, A.V. Sokolov: Dokl. Adad. NAUK SSSR, 76 (2), 197 (1951)
     S.V. Wonsowski, A.V. Sokolov, A.S. Wexler: Fortschr. Phys. 4, 216 (1956)
4.58 J. Hölzl, G. Porsch: Thin Solid Films 28, 93 (1975)
4.59 K. Christmann, G. Ertl, O. Schober: Z. Naturf. 29a, 1516 (1974)
4.60 P. Oelhafen: Thesis ETH Zürich (1976);
     P. Oelhafen, E. Gisler, F. Greuter, U. Gubler, H.P. Preiswerk; Private Commun.
     (1977)
4.60a V. Fomenko: Handbook of Thermionic Properties (Plenum, New York 1966)
4.61 A.H. Cottrell, S.C. Hunter, F.R.N. Nabarro: Phil. Mag. 44, 1064 (1953)
4.62 C. Herring: Phys. Rev. 171, 1361 (1968)
4.63 A.J. Dessler, F.C. Michel, H.E. Rorschach, G.T. Tramell: Phys. Rev. 168, 737
     (1968)
4.64 R.I. Mints, V.P. Melekhin, M.B. Partenskii: Fiz. Tverd. Tela 16, 3584 (1974)
     (English translation: Sov. Phys. Sol. State 16, 2330 (1975))
4.65 J.W. Beams: Phys. Rev. Letters 21, 1093 (1968)
4.66 P.P. Craig: Phys. Rev. Letters 22, 700 (1969)
4.67 P.P. Craig, V. Radeka: Rev. Sci. Instr. 41, 258 (1970)
4.68 S.H. French, J.W. Beams: Phys. Rev. B1, 3300 (1970)
4.69 M. Cohen, Y. Goldstein, B. Abeles: Phys. Rev. B3, 2223 (1971)
4.70 L.I. Schiff, M.V. Barnhill: Phys. Rev. 151, 1067 (1966)
4.71 K.H. Leners, R.J. Kearney, M.J. Dresser: Phys. Rev. B6, 2943 (1972)
4.72 F.C. Witteborn, W.M. Fairbank: Phys. Rev. Letters 19, 1049 (1967)
4.73 J. Strnad: Contemp. Phys. 12, 187 (1971)
4.74 C.R. Brown, J.B. Browne, E. Enga, M.R. Halse: J. Phys. D 4, 298 (1971)
4.75 J.C. Schumacher, W.E. Spicer, W.A. Tiller: Bull Am. Phys. Soc. 17, 134 (1972)
4.76 T.J. Rieger: Phys. Rev. B2, 825 (1970)
4.77 S. Berge, P.O. Gartland, B.J. Slagsvold: Surf. Sci. 48, 275 (1974)
4.78 G. Bergeret, M. Abon, B. Tardy, S.J. Teichner: J. Vac. Sci. Technol. 11, 1193
     (1974)
4.79 B.G. Baker, B.B. Johnson, G.L.C. Maire: Surf. Sci. 24, 572 (1971)

4.80 P.N. Chistyakov, R.A. Milovanova, V.V. Lytkin: Sov. Phys. Techn. phys. 19, 148 (1974)
4.81 T.A. Callcott, A.U. Mac Rae: Phys. Rev. 178, 966 (1969)
4.82 R.M. Eastman, C.H.B. Mee: J. Phys. F. Metal Phys. 3, 1738 (1973)
4.83 D.E. Eastman: Phys. Rev. B, 2, 1 (1970)
4.84 P.O. Gartland, S. Berge, B.J. Slagsvold: Phys. Rev. Lett. 28, 738 (1972)
4.85 G.A. Haas, R.E. Thomas: J. Appl. Phys. 40, 3919 (1969)
4.86 H. Kobayashi, S. Kato: Surf. Sci. 18, 341 (1969)
4.87 J. Wysocki: Acta Phys. Pol. A 39, 153 (1971)
4.88 S. Hellwig, J.H. Block: Z. Phys. Chemie 83, 269 (1973)
4.89 A. Kashetov, N.A. Gorbati: Sov. Phys. Solid State 10, 1673 (1969); translated from Fizika Tverdogo Tela 10, 2135 (1968)
4.90 R.P. Leblanc, B.C. Vanbrugghe, F.E. Girouard: Canadian Journ. Phys. 52, 1589 (1974)
4.91 R.W. Strayer, W. Mackie, L.W. Swanson: Surf. Sci. 34, 225 (1973)
4.92 B.E. Nieuwenhuys, W.H.M. Sachtler: Surf. Sci. 34, 317 (1973)
4.93 P.E.C. Franken, V. Ponec: Surf. Sci. 53, 341 (1975)
4,94 D.M. Collins, J.B. Lee, W.E. Spicer: Surf. Sci. 55, 389 (1976)
4.95 R. Bouwman, W.H.M. Sachtler: Surf. Sci. 24, 350 (1971); R. Bouwman: Thesis University Leyden (1970)
4.96 P.E.C. Franken, R. Bouwman, B.E. Nieuwenhuys, W.H.M. Sachtler: Thin Solid Films 20, 243 (1974)
4.97 T.V. Vorburger, D. Pehn, E.W. Plummer: Surf. Sci. 48, 417 (1975)
4.98 T.W. Hall, C.H.B. Mee: Phys. Stat. Sol. (a) 21, 109 (1974)
4.99 R.L. Ramey, S.J. Katzberg: J. Chem. Phys. 53, 1347 (1970)
4.100 R.J. D'Arcy, N.A. Surplice: Surf. Sci. 36, 783 (1973)
4.101 G. Haufler, H. Goretski, J. Jucker: 3rd Internat. Conf. on Thermionic Electrical Power Generation, Jülich, Germany
4.102 Ts.S. Marinova, Yu.V. Zubenko: Fiz. Tverd. Tela 12, 516 (1970), Engl. transl.: Sov. Phys. Solid State 12, 398 (1970)
4.103 O.D. Protopopov, E.V. Mikheeva, B.N. Sheinberg, G.N. Schuppe: Sov. Phys. Solid State 8, 909 (1966)
4.104 L.P. Mosteller, T. Huen, F. Wooten: Phys. Rev. 184, 364 (1969)
4.105 Kh.B. Khokonov, S.M. Zadumkin, B.B. Alchagirov: Elektrokhimiya 10, 911 (1974)
4.106 V.K. Medvedev, T.P. Smereka: Sov. Phys. Sol. State 15, 507 (1973); Transl.: Fiz. Tverd. Tela 15, 724 (1973)
4.107 H.E. Albrecht: Thesis University Rostock (1970)
4.108 L. Gaudart, R. Rivoira: C.R. Acad. Sci. Ser. B 272, 855 (1971)
4.109 K.A. Kress, G.J. Lapeyre: Sol. State Commun. 9, 827 (1971)
4.110 R. Blaszczyn, M. Blaszczyn, R. Meclewski: Surf. Sci. 51, 396 (1975)
4.111 D.L. Fehrs, R.E. Stickney: Surf. Sci. 24, 309 (1971)
4.112 T.W. Hall, C.H.B. Mee: Proc. 2nd Intern. Conf. on Solid Surfaces (1974) Japan. J. Appl. Phys. Suppl. 2, Pt. 2 (1974)
4.113 A.P. Ovchinikov, B.M. Tsarev: Sov. Phys. Sol. State 9, 2766 (1968)
4.114 W. Eib: unpublished
4.115 S. Trasatti: Surf. Sci. 32, 735 (1972)
4.116 B.Y. Chao, F.A. White: Intern. Jour. Mass Spectrom. and Ion Phys. 12, 423 (1973)
4.117 J. Polanski, Z. Sidorski, S. Zuber: Acta Physica Polonica A 49, 299 (1976)
4.118 P.A. Anderson: Phys. Rev. 98, 1739 (1955)
4.119 G.R. Garron: Compt. Rend. 258, 1458 (1964)
4.120 G.J.M. van der Velder: cited in Philips Tech. Rev. 24, 228 (1962/63)
4.121 F. Meier: Phys. Rev. B15, 4537 (1977)
4.122 J.C. Simmons: Phys. Rev. Letters 10, 10 (1963)
4.123 E.M. Savitskiy, V.F. Terekhova, E.V. Maslova: Radio Engr. Electron. Phys. 7, 1233 (1967)
4.124 K. Ueda, R. Shimizu: Japan J. Appl. Phys. 11, 916 (1972)
4.125 A.W. Dweydari, C.H.B. Mee: Phys. Stat. Sol. A 27, 223 (1975)
4.126 A.W. Dweydari, C.H.B. Mee: Phys. Stat. Sol. A 17, 247 (1973)
4.127 H.C. Potter, J.M. Blakeley: J. Vac. Sci. Technol. 12, 635 (1975) H.C. Potter, Ph.D. Thesis (Cornell University (1970) (unpublished)

4.128 T. Gustafsson, G. Broden, O.O. Nilsson: J. Phys. F4, 2351 (1974)
4.129 F. Meier and P. Zürcher: Helv. Phys. Acta, to be published
4.130 B.E. Nieuwenhuys, R. Bouwman, W.H.M. Sachtler: Thin Sol. Films 21, 51 (1974)
4.131 D.A. Gorodetskii, A.A. Yas'ko: Sov. Phys. Sol. Stat. 13, 2928 (1972)
4.132 D.A. Gorodetskii, A.A. Yas'ko: Sov. Phys. Sol. Stat. 13, 1085 (1971)
4.133 Th.G.J. van Oirschot, M. van der Brink, W.H.M. Sachtler: Surf. Sci. 29, 189 (1972)
5.1   R.L. Park: "Chemical Analysis of Surfaces"; in Surface Physics of Materials Vol. II, J.M. Blakely Ed. Academic Press, New York (1975)
5.2   C. Herring, M.H. Nichols: Rev. Mod. Phys. 21, 185 (1949), R.V. Culver, F.C. Tompkins: Advan. Catalysis 11, 68 (1959)
5.3   T. Engel, G. Ertl: to be published
5.4   K. Christmann, O. Schober, M. Neumann: J. Chem. Phys. 60, 4528 (1974)
5.5   K. Christmann, G. Ertl and T. Pignet: Surf. Sci. 54, 365 (1976) and K. Christmann, G. Ertl: Surf. Sci. 60, 365 (1976)
5.6   E. Bauer, H. Poppa, Viswanath: Surf. Sci. 58, 517 (1976)
5.7   J.C. Tracy and P.W. Palmberg, J. Chem. Phys. 51, 4852 (1969)
5.8   R. Gomer, J.K. Holm: J. Chem. Phys. 27, 1363 (1957)
5.9   R. Gomer, W. Worthmann, R. Lundy: J. Chem. Phys. 26, 1147 (1957)
5.10 R. Gomer: Disc. Faraday Soc. 28, 23 (1959)
5.11 R. Butz, H. Wagner: Surf. Sci. 63, 448 (1977)
5.12 C. Matano: Japan J. Phys. 8, 109 (1933)
5.13 A.G. Fedorus, A.G. Naumovets, Yu.S. Vedula: Phys. Stat. Sol. A 13, 445 (1972)
5.14 Yu.S. Vedula, A.G. Naumovets: in Poverkhnostnaya diffuziya i rastekanie Izd. Nauka, Moskva (1969) p. 149
5.15 R.W. Gurney: Phys. Rev. 47, 479 (1935)
5.16 S. Andersson, U. Jostell: Surf. Sci. 46, 625 (1974)
5.17 E.P. Gyftopoulos, D. Steiner: Report 27th Annual Confer. on Physical Electronics, Cambridge, Mass., (1967) 160
5.18 J.W. Gadzuk: The Structure and Chemistry of Solid Surfaces, ed. by G.A. Somorjai (Wiley, New York 1969) p. 43-1
5.19 D.M. Newns: Phys. Rev. 178, 1123 (1969)
      P.W. Anderson: Phys. Rev. 12, 41 (1961)
5.20 R.L. Gerlach, T.N. Rhodin: Surf. Sci. 19, 403 (1970)
5.21 J. Topping: Proc. Roy. Soc. London A 114, 67 (1927)
5.22 L.D. Schmidt, R. Gomer: J. Chem. Phys. 45, 1605 (1966)
5.23 E.W. Plummer, R.D. Young: Phys. Rev. B 1, 2088 (1970)
5.24 J.W. Gadzuk: Phys. Rev. B 1, 2110 (1970)
5.25 H.D. Hagstrum, G.E. Becker: Phys. Rev. Lett. 22, 1054 (1969)
5.26 W. Schröder, J. Hölzl: Solid State Comm. 24, 777 (1977)
5.27 J.P. Muscat, D.M. Newns: J. Phys. C Solid State Phys. 7, 2630 (1974)
5.28 G. Meister, W. Malzfeldt, J. Hölzl in: Conf. DPG, Munich, March 1978
5.29 D.L. Adams, L.H. Germer: Surf. Sci. 27, 21 (1971)
5.30 D.L. Fehrs, R.E. Stickney: Surf. Sci. 24, 309 (1971)
5.31 E. Bauer, H. Poppa, G. Todd: Thin Solid Films 28, 19 (1975)
5.32 E. Bauer, H. Poppa, G. Todd, F. Bonczek: J. Appl. Phys. 45, 5164 (1974)
5.33 R. Smoluchowski: Phys. Rev. 60, 661 (1941)
5.34 H.M. Kramer, E. Bauer: to be published 1978
5.35 V.K. Agarwala, T. Fort, Jr.: Surf. Sci. 48, 527 (1975)
5.36 J.R. Anderson, N. Thompson: Surf. Sci. 28, 84 (1971)
5.37 D.L. Adams, L.H. Germer, J-W. May: Surf. Sci. 22, 45 (1970)
5.38 V.K. Agarwala, T. Fort, Jr.: Surf. Sci. 45, 470 (1974)
5.39 J.R. Anderson, N. Thompson: Surf. Sci. 26, 397 (1971)
5.40 S. Andersson, U. Jostell: Sol. State Comm. 13, 829 (1973)
5.41 D.L. Adams, L.H. Germer: Surf. Sci. 32, 205 (1972)
5.42 P.E. Gregory, P. Chye, H. Sunami, W.E. Spicer: J. Appl. Phys. 46, 3525 (1975)
5.43 R.L. Wells, T. Fort, Jr.: Surf. Sci. 32, 554 (1972)
5.44 B.A. Boiko, D.A. Gorodetskii, A.A. Yas'ko: Sov. Phys. Sol. State 15, 2101 (1974) Fiz. Tverd. Tela 15, 3145 (1973)
5.45 K.H. Besocke: Materials of the 3rd Internat. Conf. on Thermionic Electrical Power Generation, Jülich, Germany (1972)

148

5.46 P.L. Young, R. Gomer: Surf. Sci. 44, 268 (1974)
5.47 M. Bacal, J.L. Desplat, T. Alleau: J. Vac. Sci. Techn. 9, 851 (1972)
5.48 Ch.S. Bhatia, M.K. Sinha: Surf. Sci. 43, 369 (1974)
5.49 R.A. Collins, B.H. Blott: J. Phys. D: Appl. Phys. 4, 102 (1971)
5.50 R.A. Collins: Surf. Sci. 26, 624 (1971)
5.51 A. Cetronio, J.P. Jones: Surf. Sci. 44, 109 (1974)
5.52 K. Christmann, O. Schober, G. Ertl: J. Chem. Phys. 60, 4719 (1974)
5.53 E. Chrzanowski: Acta Phys. Polon. A 44 , 711 (1973)
5.54 T.A. Delchar: Surf. Sci. 27, 11 (1971)
5.55 A.W. Dweydari, C.B.H. Mee: Phys. Stat. Sol. (a) 17, 247 (1973)
5.56 M.J. Dresser, Th.E. Madey, J.T. Yates, Jr.: Surf. Sci. 42, 533 (1974)
5.57 T. Engel, H. Niehus, E. Bauer: Surf. Sci. 52, 237 (1975)
5.58 B.M. Zykov, D.S. Ikonnikov, V.K. Tskhakaya: Sov. Phys. Sol. State 17, 163 (1975)
5.59 D.A. Gorodetskii, A.N. Knysh: Surf. Sci. 40, 636 (1973)
5.60 G.H. Hall, C.H.B. Mee: Phys. Stat. Sol. (a) 12, 509 (1972)
5.61 S. Hellwig, J.H. Block: Surf. Sci. 29, 523 (1972)
5.62 L. Hilaire, L. Whalley: Surf. Sci. 32, 253 (1972)
5.63 B.J. Hopkins, G.R. Shah: Vacuum 22, 267 (1972)
5.64 W. Jaschinski, R. Niedermayer: Thin Sol. Films 32, 181 (1976)
5.65 C.W. Jowett, B.J. Hopkins: Surf. Sci. 22, 392 (1970)
5.66 J. Küppers: Vacuum 21, 393 (1971)
5.67 O.V. Kanash, A.G. Naumovets, A.G. Fedorus: Sov. Phys. JETP 40, 903 (1975), ZH. Eksp. Teor. Fiz. 67, 1818 (1974)
5.68 O.K. Kultashev, A.P. Makarov: Izv. Akad. Nauk SSSR Sov. Fiz. 35, 351 (1971)
5.69 C.A. Kiwanga, R.A. Collins: Phys. Stat. Sol. (a) 23, 209 (1974)
5.70 V.K. Medvedev, A.G. Naumovets, T.P. Smereka: Surf. Sci. 34, 368 (1973)
5.71 Th.E. Madey, H.A. Engelhardt, D. Menzel: Surf. Sci. 48, 304 (1975)
5.72 V.K. Medvedev, T.P. Smereka: Sov. Phys. Sol. Stat. 15, 507 (1973), Fiz. Tverd. Tela 15, 724 (1973)
5.73 V.K. Medvedev, T.P. Smereka: Sov. Phys. Sol. Stat. 15, 1106 (1973) Fiz. Tverd. Tela 16, 1641 (1973)
5.74 V.K. Medvedev, T.P. Smereka: Sov. Phys. Sol. Stat. 16, 1046 (1974) Fiz. Tverd. Tela 16, 1599 (1974)
5.75 V.K. Medvedev, A.I. Yakivchuk: Sov. Phys. Sol. Stat. 16, 634 (1974) Fiz. Tverd. Tela 16, 981 (1974)
5.76 V.K. Medvedev, A.I. Yakivchuk: Sov. Phys. Sol. Stat. 17, 7 (1975) Fiz. Tverd. Tela 17, (1975)
5.77 C.A. Papageorgopoulos, J.M. Chen: Sol. Stat. Comm. 13, 1455 (1973)
5.78 J.H. Pollard: Surf. Sci. 20, 269 (1970)
5.79 C.A. Papageorgopoulos, J.M. Chen: Surf. Sci. 52, 40 (1975)
5.80 P.W. Palmberg: Surf. Sci. 25, 598 (1971)
5.81 C.A. Papageorgopoulos, J.M. Chen: Surf. Sci. 39, 283 (1973)
5.82 T.N. Taylor, P.J. Estrup: J. Vac. Sci. Technol. 11, 244 (1974)
5.83 Yu.S. Vedula, Yu.M. Konoplev, V.K. Medvedev, A.G. Naumovets, T.P. Smereka, A.G. Fedorus: Materials of the 3rd Internat. Conference on Thermionic Electrical Power Generation, Jülich, Germany 1972
5.84 G. Wedler, H. Papp, G. Schroll: Surf. Sci. 44, 463 (1974)
5.85 J.L. Wells, T. Fort, Jr.: Surf. Sci. 33, 172 (1972)
5.86 L.F. Wagner, W.E. Spicer: Surf. Sci. 46, 301 (1974)
5.87 R. Blaszczyszyn, M. Blaszczyszyn, R. Meclewski: Surf. Sci. 51, 396 (1975)
5.88 J.T. Yates, Jr., Th.E. Madey: J. Chem. Phys. 54, 4969 (1971)
5.89 Z. Sidorski, I. Pelly, R. Gomer: J. Chem. Phys. 50, 2382 (1969)
5.90 T. Fort, Jr., R.L. Wells: Surf. Sci. 32, 543 (1972)
5.91 V.K. Medvedev, A.G. Naumovets, A.G. Fedorus: Sov. Phys. Sol. Stat. 12, 301 (1970) Fiz. Tverd. Tela 12, 375 (1970)
5.92 B.M. Zykov, D.S. Ikonnikov, V.K. Tskhakaya: Sov. Phys. Sol. Stat. 17, 2322 (1976) Fiz. Tverd. Tela 17, 3562 (1975)
5.93 Yu.V. Zubenko: Sov. Phys. Sol. Stat. 14, 377 (1972) Fiz. Tverd. Tela 14, 454 (1972)
5.94 D.A. Gorodetskii, Yu.P. Mel'nik: Izv. Akad. Nauk SSSR Ser. Fiz. 35, 1064 (1971)

5.95 J. Zebrowski: Acta Phys. Polon. A44, 201 (1973)
5.96 E.V. Klimenko, V.K. Medvedev: Sov. Phys. Sol. Stat. 10, 1562 (1969) Fiz. Tverd. Tela 10, 1986 (1968)
5.97 D.A. Gorodetskii, Yu.P. Mel'nik: Sov. Phys. Sol. Stat. 16, 1805 (1975) Fiz. Tverd. Tela 16, 2781 (1974)
5.98 J.M. Chen, C.A. Papageorgopoulos: Sol. Stat. Somm. 11, 999 (1972)
5.99 H. Conrad, G. Ertl, J. Koch, E.E. Latta: Surf. Sci. 43, 462 (1974)
5.100 G.E. Becker, H.D. Hagstrum: Surf. Sci. 30, 505 (1972)
5.101 B.D. Barford, R.R. Rye: J. Chem. Phys. 60, 1046 (1974)
5.102 J.M. Chen, C.A. Papageorgopoulos: Surf. Sci. 26, 499 (1971)
5.103 H.A. Engelhardt, D. Menzel: Surf. Sci. 57, 591 (1976)
5.104 N.R. Avery: Surf. Sci. 43, 101 (1974)
5.105 G. Dalmai-Imelik, J.C. Bertolini: Proc. 2nd Internat. Conf. on Solid Surfaces, (1974) Japan. J. Appl. Phys. Suppl. 2, Pt. 2, 177 (1974)
5.106 E. Bauer, H. Poppa, G. Todd, P.R. Davis: J. Appl. Phys. 48, 3773 (1977)
5.107 G. Dalmai-Imelik, J.C. Bertolini: Proc. 2nd Intern. Conf. on Sol. Surf. 1974 Japan. J. Appl. Phys. Suppl. 2, Pt. 2, 205 (1974)
5.108 J.E. Demuth, T.N. Rhodin: Surf. Sci. 45, 249 (1974)
5.109 O.K. Kultashev, A.P. Makarov: Izv. Akad. Nauk SSSR Ser. Fiz. 38, 317 (1974)
5.110 A.P. Janssen, J.P. Jones: Surf. Sci. 41, 257 (1974)
5.111 B.J. Hopkins, C.B. Williams, P.C. Wilmer: Surf. Sci. 25, 633 (1971)
5.112 P.H. Holloway, J.B. Hudson: Surf. Sci. 43, 123 (1974)
5.113 P.H. Holloway, J.B. Hudson: Surf. Sci. 43, 141 (1974)
5.114 D.A. Gorodetskii, A.N. Knysh: Sov. Phys. Sol. Stat. 13, 2119 (1972) Fiz. Tverd. Tela 13, 2521 (1971)
5.115 J.L. Gland, G.A. Somorjai: Surf. Sci. 41, 387 (1974)
5.116 J. Fusy, B. Bigeard, A. Cassuto: Surf. Sci. 46, 177 (1974)
5.117 P.E.C. Franken, V. Ponec: Surf. Sci. 53, 341 (1975)
5.118 R. Müller, H.-W. Wassmuth: Surf. Sci. 40, 15 (1973)
5.119 Th.E. Madey: Surf. Sci. 29, 571 (1972)
5.120 A.J. Melmed, J.J. Carroll, R. Meclewski: Surf. Sci. 45, 649 (1974)
5.121 V.I. Makhov, B.V. Bondarenko: Sov. Phys. Sol. Stat. 12, 2986 (1971) Fiz. Tverd. Tela 12, 3661 (1970)
5.122 Ts.S. Marinova, Yu. and V. Zubenko: Sov. Phys. Sol. Stat. 12, 398 (1970) Fiz. Tverd. Tela 12, 516 (1970)
5.123 T.J. Lee, B.H. Blott, B.J. Hopkins: J. Phys. F: Metal Phys. 1, 309 (1971)
5.124 C. Lea, R. Gomer: J. Chem. Phys. 54, 3349 (1971)
5.125 R.A. Collins, B.H. Blott: J. Phys. D: Appl. Phys. 4, 114 (1971)
5.126 N.G. Krishnan, W.N. Delgass, W.D. Robertson: Surf. Sci. 57, 1 (1976)
5.127 K. Kraemer, D. Menzel: Berichte der Bunsengesellschaft 78, 591 (1974)
5.128 K. Kraemer, D. Menzel: Berichte der Bunsengesellschaft 78, 728 (1974)
5.129 D.F. Klemperer, J.C. Snaith: Surf. Sci. 28, 209 (1971)
5.130 E.V. Klimenko, A.G. Naumovets: Sov. Phys. Sol. Stat. 15, 2181 (1974) Fiz. Tverd. Tela 15, 3272 (1973)
5.131 G. Rovida, F. Pratesi, M. Maglietta, E. Ferroni: Surf. Sci. 43, 230 (1974)
5.132 R. Riwan, C. Guillot, J. Paigne: Surf. Sci. 47, 183 (1975)
5.133 T. Radon: Acta Phys. Polon. A43, 699 (1973)
5.134 C.A. Papageorgopoulos, J.M. Chen: J. Vac. Sci. and Technol. 9, 570 (1971)
5.135 C.A. Papageorgopoulos, J.M. Chen: Surf. Sci. 39, 313 (1973)
5.136 C.A. Papageorgopoulos, J.M. Chen: Surf. Sci. 52, 53 (1975)
5.137 J. Polanski, Z. Sidorski: Surf. Sci. 40, 282 (1973)
5.138 B.M. Palyukh, T.P. Smereka: Sov. Phys. Sol. Stat. 13, 640 (1971) Fiz. Tverd. Tela 13, 776 (1971)
5.139 B.E. Nieuwenhuys, R. Bouwman, W.M.H. Sachtler: Thin Solid Films 21, 51 (1974)
5.140 A. Mroz, Z. Sidorski: Acta Phys. Polon. A49, 437 (1976)
5.141 J. Workowski: Acta Phys. Polon. A42, 9 (1972)
5.142 R.J. Whitefield, J.J. Brady: Phys. Rev. Lett. 26, 380 (1971)
5.143 N.V. Volkov, Yu. and K. Gus'kov, U.N. Kononova, Yu.I. Kostikov: Sov. Phys. Tech. Phys. 19, 144 (1974) Zh. Tekh. Fiz. 44, 224 (1974)
5.144 J.C. Tracy: J. Chem. Phys. 56, 2736 (1972)
5.145 N.A. Surplice, W. Brearley: Surf. Sci. 52, 62 (1975)
5.146 H.W. Wassmuth: Annalen d. Physik 26, 326 (1971)

5.147 J.C. Tracy: J. Chem. Phys. 56, 2748 (1972)
5.148 H. Conrad, G. Ertl, E.E. Latta: J. of Catalysis 35, 363 (1974)
5.149 H. Conrad, G. Ertl, E.E. Latta: Surf. Sci. 41, 435 (1974)
5.150 B.G. Baker, B.B. Johnson: J. of Vac. Sci. and Technol. 9, 930 (1971)
5.151 J. Polanski, Z. Sidorski, S. Zuber: Acta Phys. Polon. A49, 299 (1976)
5.152 J.T. Yates, Jr., Th.E. Madey: Surf. Sci. 28, 437 (1971)
5.153 D.L. Fehrs: Surf. Sci. 17, 298 (1969)
5.154 N.I. Ionov, B.K. Medvedev: Sov. Phys. Sol. Stat. 16, 1719 (1975) Fiz.
      Tverd. Tela 16, 2651 (1974)
6.1   T.E. Fisher: J. Vac. Sci. Techn. 11, 252 (1974)
6.2   R.L. Moss, L. Whally: Advanc. Catal. 22, 115 (1972); T.E. Fisher: J. Vac.
      Sci. Techn. 11, 252 (1974); V. Ponec: Catal. Rev. Sci. Eng. 11, 1 (1975)
6.3   R. Bouwman, G.J.M. Lippits, W.M.H. Sachtler: J. Catal. 25, 350 (1972);
      S.H. Overburg, G.A. Somorjai: Surf. Sci. 55, 209 (1976)
6.4   N.F. Mott: Phil. Mag. 22, 287 (1936)
6.5   G. Ertl, K. Wandelt: Phys. Rev. Lett. 29, 218 (1972)
6.6   D.E. Eastman: "Electron Spectroscopy"; in Proc. of the Int. Conf., Asilomar
      (Cal.) (1971), North Holland, Amsterdam (1972), p. 487
6.7   P. Soven: Rev. 178, 1136 (1969
      B. Velicky, S. Kirckpatrick, H. Ehrenreich: Phys. Rev. 175, 747 (1968)
6.8   C.D. Gelatt, Jr., H. Ehrenreich: Phys. Rev. B 10, 398 (1974)
6.9   S.C. Fain, Jr., J.M. McDavid: Phys. Rev. 139, 5099 (1974)
6.10 Handbook of Thin Film Technology, L.I. Maissel, R. Glang edts. McGraw-Hill,
      New York 1970
6.11 B.Ch. Dyubua, O.K. Kultashev, L.V. Gorshkova: Fiz. Tverd. Tela 8, 1105 (1966),
      Engl. transl.: Sov. Phys. Solid State 8, 882 (1966)
6.12 R. Bouwman: Thesis University Leyden (1970)
6.13 R. Bouwman, G.J.M. Lippits, W.M.H. Sachtler: J. Catal. 25, 350 (1972)
6.14 W.M.H. Sachtler: J. Vac. Sci. Technol. 9, 828 (1971)
6.15 F.L. Williams, D. Nason: Surf. Sci. 45, 377 (1974)
6.16 W.M.H. Sachtler, G.J.H. Dorgelo: J. Catal. 4, 654 (1965)
6.17 V. Ponec: Catal. Rev. Sci. Eng. 11 (1), 1 (1975)
6.18 K. Christmann, G. Ertl: Surf. Sci. 33, 254 (1972)
6.19 H.H. Brongersma, T.M. Buck: Surf. Sci. 53, 649 (1975)
6.20 G. Ertl, J. Küppers: Surf. Sci. 24, 104 (1971)
6.21 Y. Takasu, H. Konno, T. Yamashina: Surf. Sci. 45, 321 (1974)
6.22 C.R. Helms: J. Catal. 36, 114 (1975);
      C.R. Helms, K.Y. Yu: J. Vac. Sci. Techn. 12, 276 (1975)
6.23 S.C. Fain, Jr., J.M. McDavid: Phys. Rev. B9, 5099 (1974)
6.24 S.C. Fain, Jr., J.M. McDavid: Phys. Rev. B13, 1853 (1976)
6.25 P.E.C. Franken, V. Ponec: J. Catal. 42, 398 (1976)
6.26 E.R. Davidson, S.C. Fain, Jr.: J. Vac. Sci. Technol. 13, 209 (1976)
6.27 Ju.I. Malov, M.D. Shebzukov, V.B. Lazarev: Surf. Sci. 44, 21 (1974)
6.28 H.C. Potter, J.M. Blakely: J. Vac. Sci. Technol. 12, 635 (1975)
6.29 P. Smoluchowski: Phys. Rev. 60, 661 (1941)
6.30 J.P. Appelbaum, D.R. Hamann: Phys. Rev. Lett. 31, 1106 (1973)
6.31 W.J. Müller, M. Freissler, E. Plettinger: Z. Electrochem. 42, 366 (1936)
6.32 T.C. Tisone, J. Dobrek: J. Vac. Sci. Technol. 1, 271 (1972)
6.33 E.M. Hörl: J. Vac. Sci. Technol. 1, 276 (1972)
6.34 W.B. Novak, R.N. Dyer: J. Vac. Sci. Technol. 1, 279 (1972)
6.35 J.R. Rairden, C.A. Neugebauer, R.A. Sigsbee: General Electric Research and
      Evelopment Center, Schenectady, New York Report No. 70-C-235 (1970)
      (unpublished)
6.36 R.E. Thomas, G.A. Haas: J. Appl. Phys. 43, 4900 (1972)
6.37 R.E. Thomas: J. Appl. Phys. 41, 5330 (1970)

Physical and Chemical Properties of Stepped Surfaces

H. Wagner

1. Introduction

Within the last two decades, fundamental research on solid surfaces has been mainly concerned with polycrystalline films and low index planes of single crystals,
their physical properties, and interaction with foreign molecules. The availability of single-crystal specimens of a variety of metallic, semiconducting, and in-
sulating elements or compounds and the routine application of ultrahigh vacuum
conditions formed the basis for investigating well-defined surfaces. In addition,
surface sensitive experimental methods such as low energy electron diffraction
(LEED) and Auger electron spectroscony (AES) became routinely used to characterize
structure and chemical composition or cleanliness of the surface under investigation. A wealth of further experimental tools tailored to study specific surface
properties helped to unravel the intriguing and often rather complex processes
occurring on solid surfaces. However, even in the case of well-characterized low
index planes, a complete understanding of certain properties and interactions with
foreign molecules or atoms has sometimes proven to be difficult.

   Surface cleanliness, one prerequisite for systematic and proper surface studies,
has posed the biggest problem in the past but now seems to be well unter control
(in most cases impurities < 1 %) by using suitable cleaning procedures checked by
AES. The microscopic surface structure which is likewise important to be known is
usually considered to comply pretty well with the ideal atom arrangement of the
respective low index planes. Sharp diffraction patterns observed in LEED tempt one
to overemphasize the crystalline perfection of these surfaces. The diffraction
features, however, are rather insensitive to randomly distributed surface defects
such as adatoms, vacancies, and steps. It is intuitively expected and, in some
cases (e.g., crystal growth), has been known for a long time that surface defects
may alter the physical properties of crystal planes and influence their behavior
with respect to interactions with foreign molecules (adsorption, desorption, chemical reactions, etc.). This influence of surface defects was recently reviewed
by RHEAD /1/.

   In order to study the influence of defects, it is desirable to have surfaces
available with defined defects of known concentration. Surface atoms in edge and

kink positions represent the largest portion of possible defects on crystalline surfaces. Vicinal surfaces, i.e., surfaces with an orientation close (vicinal) to a low index orientation, provide in a natural way for atomic steps of known direction and quantity. The number of steps is simply related to the inclination angle towards the low index plane. The step direction given by the intersection of the vicinal plane with the low index plane determines the density of atoms in kink positions. The steps are supposed to be equally spaced which results in a periodic array of low index terraces separated by monoatomic steps. We shall refer to vicinal surfaces exhibiting this structural feature as regularly stepped surfaces. HENZLER /2/ described methods (especially LEED) for studying steps on single crystals and summarized some important results.

This article is devoted to various aspects of regularly stepped surfaces. In the following section we shall present experimental evidence for periodic step structures on vicinal surfaces as obtained by field ion microscopy (FIM), LEED, and electron microscopy. The various methods are briefly described, and special emphasis is given to the characterization of periodic step arrays by LEED because this method has been widely used in the field.

The third section treats physical properties of clean stepped surfaces. One problem of major concern involves the question whether the state of lowest surface energy of a vicinal surface is represented by a periodic step structure or rather by a possible hill and valley structure formed by other low index facets. This question is intimately related to the thermal stability of regular step structures. Experimental observations show that both structural features are possible, depending on the specific material and kind of vicinal surface. Theoretical considerations in this respect are qualitatively tractable but do not allow for definite statements in the specific case because the pertinent input parameters are unknown or difficult to determine. Electronic properties of stepped surfaces have been treated experimentally as far as the work function is concerned and both theoretically and experimentally in the case of surface states.

The fourth section deals with the interaction of stepped surfaces with foreign molecules and atoms. Especially the kinetics of adsorption and reaction processes deserve considerable attention in view of the technological importance with respect to epitaxial growth, corrosion, and, above all, heterogeneous catalysis. The concept of "active sites" in heterogeneous catalysis has often been associated with specific surface sites. Step and kink sites are places of higher coordination and may be especially active in the case of "structure sensitive" reactions /3/. One major incentive for studying stepped surfaces therefore concerns the prospect of adding a contribution to the understanding of heterogeneous catalysis.

Many of the experimental results reviewed in this article are only qualitatively understood or may be explained by plausibility arguments. The theoretical treatment of the underlying physical problems is rather complicated. The latter, however, holds true even for similar properties or processes connected with low index planes. One major objective has been to clearly demonstrate and, as far as possible, to state quantitatively the differences in physical properties and kinetic processes related to a given low index plane and the corresponding stepped surface, respectively. More systematic experimental studies are needed to elucidate the physical reasons which are responsible for the observed effects associated with steps and to develop concepts which may be treated theoretically.

2. Characterization of Stepped Surfaces

2.1 Surface Crystallography

Crystal surfaces of low index orientation, in most cases, reflect the symmetry of the bulk structure, i.e., their lateral atomic arrangement is identical to the arrangement of atoms in a bulk lattice plane of the same Miller index. The unit mesh of the two-dimensional surface structure contains one or only a few atoms whose number depends on the crystal structure and the surface orientation. These low index surfaces have relative minima in surface energy with regard to surfaces of vicinal orientations (vicinal surfaces). High temperatures close to the melting point are required to produce steps, adatoms, or vacancies on such singular surfaces /4/. What surface topography can be expected if one departs from the singular surface and moves on to a vicinal surface? One may expect /5/ that the vicinal surface is composed of terraces of the low index orientation which are separated by monoatomic steps. This model surface is an example of the terrace-ledge-kink (TLK) model first introduced and investigated by KOSSEL et al. /6/ and STRANSKI et al. /7/. The step direction and the average step separation are determined by the orientation of the vicinal surface. Fig. 2.1 shows a hard sphere model of a vicinal surface close to the (110) plane of a bcc crystal structure. Suppose the normal of the macroscopic vicinal surface is inclined to the normal of the low index surface by the angle α and the step height is d (which is equal to the separation between the low index planes). The average distance between neighboring steps is then given by

$$\Lambda = d/\tan\alpha. \tag{2.1}$$

The average direction of the steps is parallel to the line of intersection between low index and vicinal surface. The step direction may coincide with the direction of close-packed atom rows (low index direction). In this case the edges are smooth.

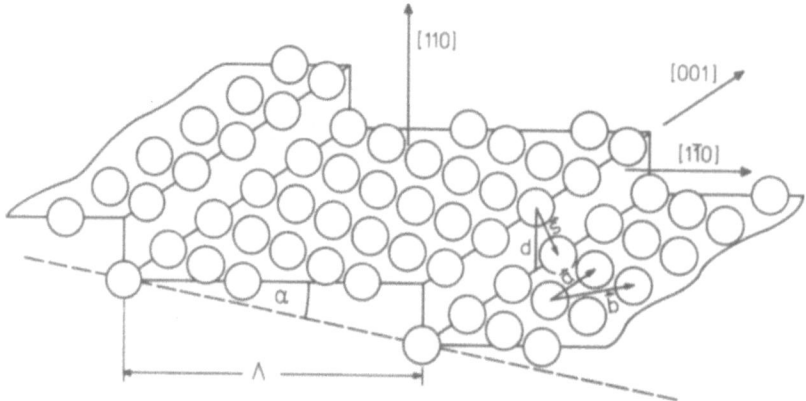

Fig. 2.1. Model of a stepped surface vicinal to the (110) plane of the bcc crystal structure, e.g., tungsten. The surface is described by W(S)-[6(110)x(1Ī0)]. Neighboring terraces are connected by the vector s. The unit mesh of the (110) plane is given by vectors a and b

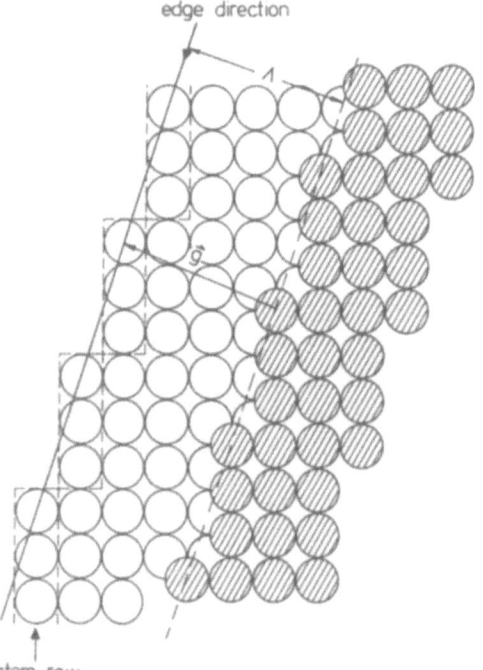

Fig. 2.2. Top view of a stepped surface indicating the definition of atom rows on the terrace. 4 2/3 atom rows are accommodated on the lower terrace (open circles). The edges are formed by smooth segments of close-packed atoms and periodically spaced kinks. The vector g connects atoms in equivalent positions on neighboring terraces

For other directions the edges will contain kinks which connect portions of smooth edges as indicated in Fig. 2.2.

For later use we shall characterize vicinal, i.e., stepped, surfaces by a nomenclature which indicates the terrace orientation, the terrace width, and the edge orientation. This nomenclature was first introduced by LANG et al. /8/. Consider again Fig. 2.1 and suppose the model surface represents a stepped (S) tungsten (W) surface with (100) terraces and each terrace accommodates on the average six atom rows. The step is one atomic layer high and the edge orientation is described by the MILLER indices (1$\bar{1}$0). The edge orientation characterizes the crystal plane which contains the edge atoms and the neighboring atoms of the lower terrace. With this nomenclature the W surface depicted in Fig. 2.1 is characterized by

$$W(S) - [6(110) \times 1(1\bar{1}0)]. \tag{2.2}$$

In general, a stepped surface is represented by

$$E(S) - [m(h,k,\ell) \times n(h',k',\ell')]. \tag{2.3}$$

E designates the element (or compound), m gives the number of rows on the terrace of (h,k,ℓ) orientation, n the number of atomic layers giving the step height, and (h',k',ℓ') the edge orientation. An additional remark should be made concerning m, the number of atom rows on a terrace. There is no ambiguity if the rows are parallel to low index directions. If the step edge follows an arbitrary direction and, therefore, contains kinks, we shall proceed as follows. All edge atoms between two neighboring equivalent edge positions, e.g., kink positions, belong to one atom row as shown in Fig. 2.2. In the general case this definition leads to fractional row numbers. Indeed, one can decrease the terrace width by taking away just the atoms located in kink positions without changing the structure and direction of the step edge. In the particular case shown in Fig. 2.2 one third of an atom row would then be taken away. This definition has the advantage that the number of atom rows on a terrace represents approximately the terrace width regardless of the respective edge direction. We shall include in the number of atom rows the row which lies in the terrace below an edge row. This row together with the row of edge atoms has already been considered to define the crystal plane which serves to characterize the edge orientation.

So far we have described the topography of vicinal surfaces in a rather ideal manner. In the next section, experimental evidence will be given which shows that indeed regular step structures are often the characteristic features of vicinal surfaces. This is the reason why vicinal surfaces are so attractive for studies of gas-solid interactions. The orientation of the vicinals simply determines the number of atoms in edge, kink, or terrace position. The terrace orientation as well

as the specific structure of an edge or kink site might also be varied. Preset
concentrations of these sites allow one to investigate the influence they might
have on processes like adsorption, desorption, catalytic reactions, nucleation,
segregation, etc..

2.2 Experimental Evidence for Step Structures

In order to investigate the topography of vicinal surfaces with inclination angles
to a low index plane of, say, 2-15°, experimental methods are required with a local resolution of the order of 10-15 Å. This is the range of step separations expected from the TLK model with monoatomic step heights. Field ion microscopy (FIM)
/9/ provides an even better resolution and allows direct imaging of steps, adatoms,
and vacancies on surfaces of field emitter tips. Low energy electron diffraction
(LEED) /10-12/ enables the study of periodic step structures with step separations
up to about 50 Å. In addition, step height determinations can be carried out with
an accuracy of 1 % even for randomly oriented steps. Electron microsopy /13-15/
also allows investigations of step structures by utilizing a suitable decoration
method /16/.

The specimens used in the different techniques are prepared according to the
special experimental requirements. For studies in the field ion microscope, thin
sample tips are required which are produced by proper etching procedures /9/ to
yield tip radii of the order of 100-1000 Å. Macroscopic sample surfaces of semiconductor (Si, Ge, GaAs) or alkali halide materials can be obtained by cleavage
/17/ This technique employed under ultrahigh vacuum conditions produces clean surfaces per se but is only applicable for certain crystallographic directions (cleavage planes). One additional drawback (sometimes, however, regarded as an experimental advantage) is the fact that only local deviations from the ideal cleavage
plane (low index orientation) form the regions of vicinal surface orientations.
The stepped regions occur more or less at random and with various inclination
angles to the low index cleavage plane. Vicinal metal surfaces are exclusively
prepared by usual metallographic procedures involving Laue back reflection for
the proper adjustment of the sample orientation, grinding, polishing, and subsequent cleaning and recrystallization of the top surface region in an ultra vacuum system /18/. Because of the difficulty to identify easily a Laue back reflection pattern of a vicinal surface it is convenient to follow an orientation
procedure using a set computed Laue back reflection plots /19/. Semiconductor surfaces of vicinal orientations have been prepared in this way as well.

2.2.1 Field Ion Microscopy

In the FIM technique a thin tip of about 100 Å to several thousand Ångstrom radius
is employed as the specimen to be investigated. Electric field strengths of the
order of several 10^8 V cm^{-1} at the tip surface cause field ionization of atoms of
an imaging gas (He, Ar, Ne). The ionized gas atoms are radially accelerated to a
fluorescent screen on which they give rise to an image of the spatial distribution
of the ionization probability on the emitter tip. Figure 2.3 is a schematic repre-

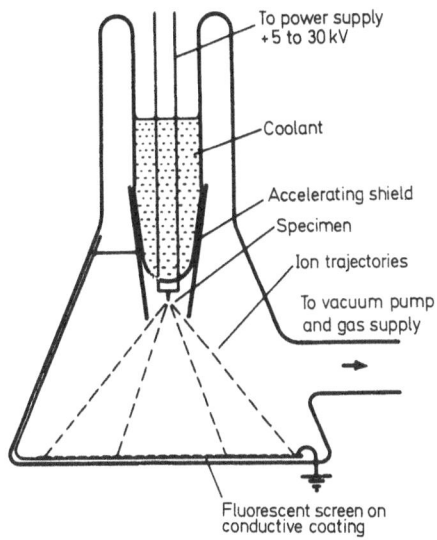

Fig. 2.3. Schematic representation of a field ion microscope (according to /9/)

sentation of a field ion microscope. The ionization probability is larger near surface atoms in protruding positions, such as adatoms, kink, or edge atoms. The image
on the fluorescent screen represents directly a large magnification (10^6) of the
tip surface topography. Figure 2.4 reproduces the FIM image of an iridium tip with
the (001) orientation in the center of the tip. The ring-shaped spot arrays around
the {001} region reflect edge and kink atoms of edges bordering (001) terraces. Similar terrace structures are visible around the other low index regions of (111)
type. Several other high index planes not marked in the figure are also surrounded
by terrace structures. The well-ordered surface structure reproduced in Fig. 2.4
was prepared by the field evaporation technique /20/. The radius of curvature close

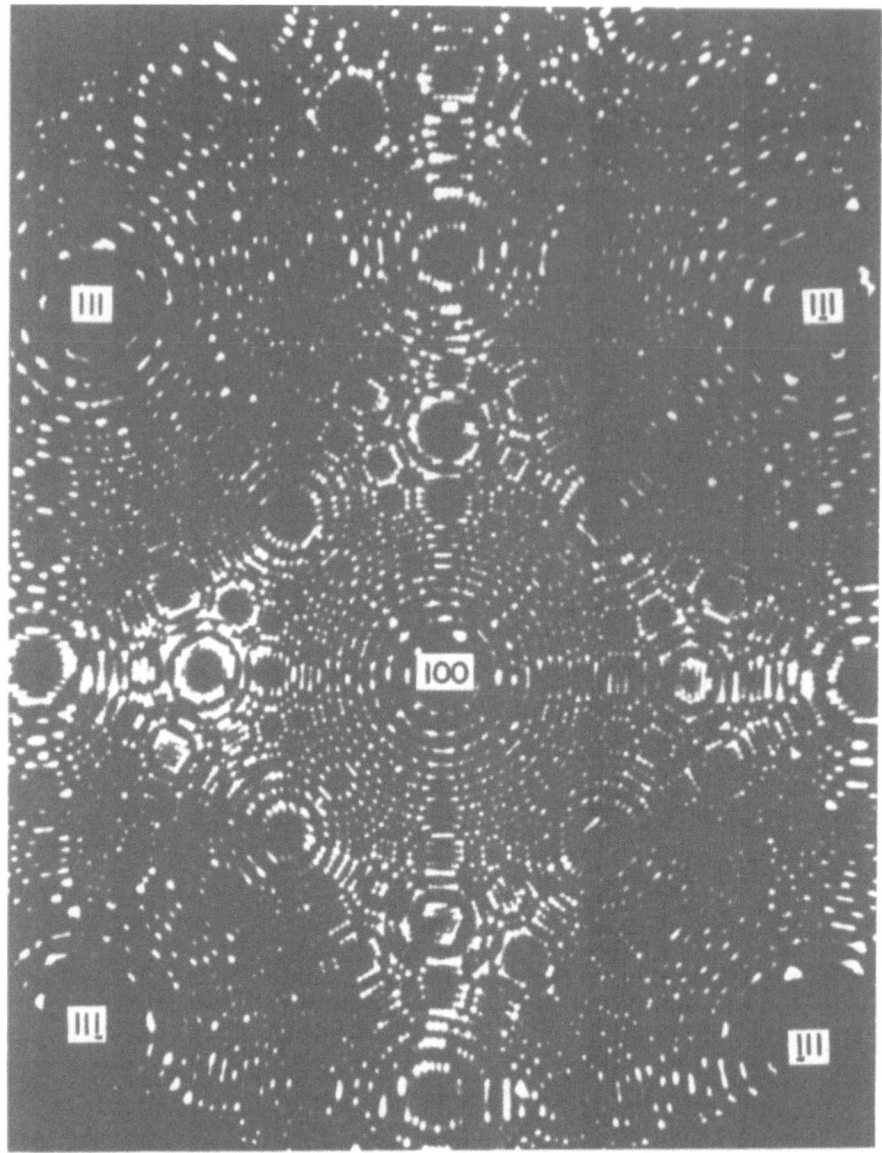

Fig. 2.4. Field ion microscope image of an iridium tip with the (100) orientation in central position (according to /9/)

to the low index planes may be obtained directly by the known monoatomic step height and the number of rings within an angular distance from the low index direction /21/.

The FIM method which was very briefly sketched in this section demonstrates that the surface structure which one would expect in the framework of a TLK model is actually observed.

2.2.2 Low Energy Electron Diffraction (LEED)

Periodic structures on solid surfaces have been extensively investigated by low
energy electron diffraction. The wavelength λ of a monochromatic electron beam of
energy $E = e_0 U$ is, according to the de Broglie relation, given by $\lambda[\overset{\circ}{A}] = \sqrt{150/U[V]}$
with U the accelerating voltage. For voltages between 20 and 600 V the wavelength
covers the range between 3 and 0.5 $\overset{\circ}{A}$ and is of the order of atomic distances. Thus
diffraction patterns representing the two-dimensional reciprocal lattice of the
sample surface can be obtained from the elastically scattered electrons. A scheme
of a LEED apparatus is shown in Fig. 2.5. Electrons with wave vector $\underline{k}_0 = 2\pi\underline{k}_0/\lambda|\underline{k}_0|$

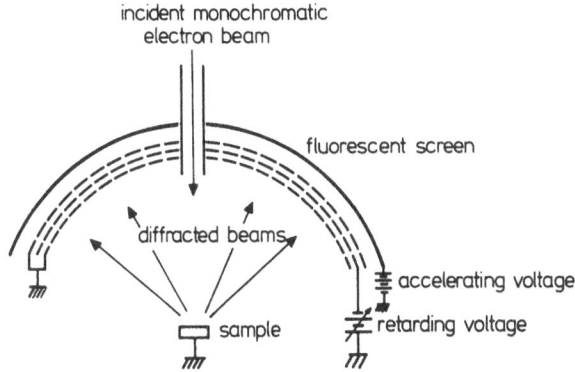

Fig. 2.5. Schematic representation of low energy electron diffraction apparatus
(LEED)

impinge on the surface, and the elastically diffracted electrons are selected by
retarding grids and post-accelerated to a fluorescent screen. The directions $\underline{k}$ of
the diffracted beams can be obtained directly from the spot pattern on the screen.
Let the two-dimensional periodic structure of the sample surface be described by
the lattice vectors $\underline{a}$ and $\underline{b}$ and the corresponding reciprocal vectors $\underline{a}^*$ and $\underline{b}^*$.
Kinematic diffraction theory states that the components of $\underline{k}$ and $\underline{k}_0$ parallel to
the sample surface differ by a reciprocal lattice vector $h\underline{a}^* + k\underline{b}^*$ with h and k
integer values. The component of $\underline{k}$ normal to the surface is determined by the con-
dition for elastic scattering $|\underline{k}| = |\underline{k}_0|$. The scattered beam wave vectors are then
labeled $\underline{k}_{h,k}$ according to the reciprocal lattice vectors. Let us visualize these
relations by the Ewald construction in Fig. 2.6, which shows for simplicity only
the two-dimensional case. The surface structure in real space is shown in the

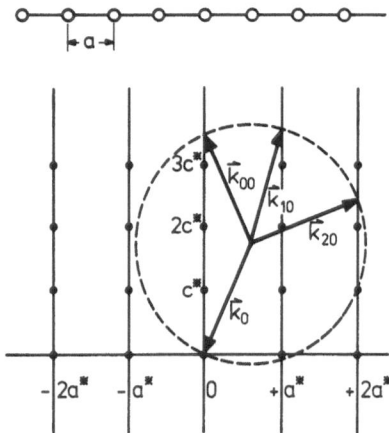

Fig. 2.6. Ewald construction for determining the directions of backscattered diffraction beams from a periodic array of surface atoms. The directions are given by the intersection of the Ewald sphere (circle) with the vertical rods erected in reciprocal lattice points

upper part of Fig. 2.6 and is represented by a chain with lattice constant a. The lower part shows the reciprocal lattice with lattice vector $\underline{a}^* = 2\pi/a$. Because of the small penetration depth of low energy electrons, only the uppermost surface planes contribute to the intensity of elastically scattered electrons. This means that the third Laue condition describing the interference between elastically scattered electrons from different lattice planes need not be fulfilled. (Nevertheless, for later use the reciprocal lattice points are indicated in the figure.) The reciprocal lattice is therefore represented by the vertical rods erected in the lattice points $h\underline{a}^*$. The wave vectors $\underline{k}_h$ of the reflected beams are obtained by the cross points of the Ewald sphere with the lattice rods. Variations of the incident wave vector $\underline{k}_o$ by changing the acceleration voltage or direction result in corresponding changes of the $\underline{k}_h$ directions and hence changes of the spot positions on the screen. Figure 2.9a shows the LEED spot pattern from a tungsten (110) face obtained at 110 V.

This brief recollection of the LEED method may suffice to appraise the LEED patterns obtained from vicinal surfaces. The first systematic LEED study of a vicinal surface was reported by ELLIS and SCHWOEBEL /22/ who prepared a uranium dioxide surface inclined 11.4° against the (111) plane. This surface corresponds closely to the (553) orientation and should, according to the TLK model, exhibit steps parallel to the [110] direction. The most striking feature of the LEED pattern from the vicinal surface consists in the appearance of double spots at distinct voltages. Each spot characteristic for the symmetry (or unit mesh) of the (111) plane changes periodically with voltage from sharp single to double spots. The characteristic voltages depend on the diffraction order (h,k).

In subsequent investigations on vicinal surfaces of Cu /23/, Re /24/, Pt /8/, Ge /25,26/, Si /27/, GaAs /28/, W /29/, Ni /30,31/, Pd /32/, Au /33,34/, and Ir /35/ the characteristic LEED features were always observed and found to be indicative of stepped surfaces. In the following section we shall discuss the variations caused by step structures in the diffraction pattern and the informations which can be extracted with regard to the step structure itself, i.e., terrace width, step height, and step orientation.

2.2.3 LEED from Stepped Surfaces

The LEED phenomena observed from vicinal surfaces have been thoroughly discussed and fully explained by the superposition of diffraction from the lattice structure of the terrace and the periodic or random step arrays. ELLIS and SCHWOEBEL /22/, RHEAD and PERDEREAU /23/, and HENZLER /17,25/ have given somewhat different descriptions of the same physical process.

Consider a stepped surface as depicted in Fig. 2.1 characterized by terraces of equal width Λ separated by monoatomic steps of height d. Let $\underline{g}$ be the vector connecting two atoms in equivalent edge positions on neighboring terraces (Fig. 2.2). The projection of $\underline{g}$ on the normal to the edge direction within the terrace plane equals the terrace width Λ. If the scattering amplitude of a single terrace is described by A, then the resulting scattering amplitude of M terraces of the step structure equals

$$A [1 + e^{i\underline{g}(\underline{k}-\underline{k}_0)} + e^{2i\underline{g}(\underline{k}-\underline{k}_0)} + \ldots + e^{(M-1)i\underline{g}(\underline{k}-\underline{k}_0)}]$$

and the scattered intensity J is given by

$$J = |A|^2 [\frac{\sin(M \psi/2)}{\sin \psi/2}]^2 \quad \text{with } \psi = \underline{g}(\underline{k}-\underline{k}_0). \tag{2.4}$$

The intensity $(A)^2$ from a single terrace has maxima in directions given by the Ewald construction of Fig. 2.6. This intensity is now modulated by the second term in (2.4) which has maxima in directions for which $\underline{g}(\underline{k}-\underline{k}_0) = 2\pi n$ with n integer. We may illustrate this finding by the two-dimensional representation /2/ of Fig. 2.7. The intensity variation $(A)^2$ is plotted versus the scattering angle ϕ. Only the positions of the maxima are being considered here. The relative heights of the maxima are not representative for LEED because of important contributions due to dynamic scattering effects. The product in (2.4) leads to the appearance of double

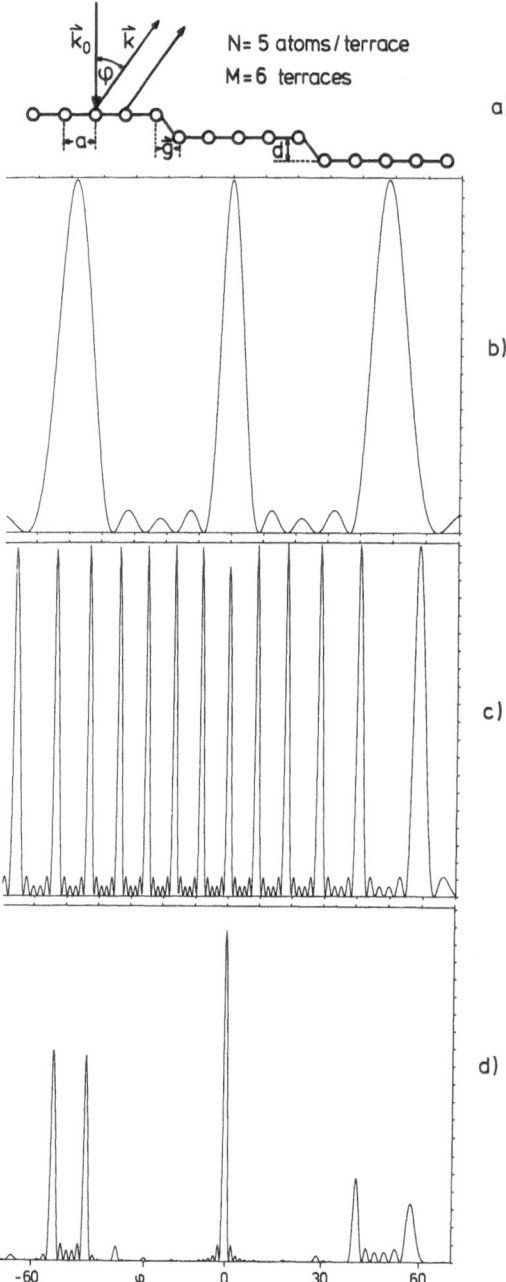

$\vec{k}_0$ $\vec{k}$

φ

N= 5 atoms/terrace

M= 6 terraces

a)

$\vdash a\dashv$ $\vdash g\dashv$ $\vdash d\dashv$

b)

c)

d)

-60 φ_s 0   30   60

Fig. 2.7. Two-dimensional representation of low energy electron diffraction from a stepped surface with periodic step array. a) Step structure in real space. Lattice constant a. b) Diffraction function from a single terrace with five atom rows as function of diffraction angle ϕ. $a/\lambda = \sqrt{9/5}$ and $d/\lambda = 3/2$. c) Diffraction function of step array with six terraces. Zero-order diffraction occurs at $\phi_s = -25.2^0$. d) Product of diffraction functions as shown in b) and c). For the wavelenth λ chosen, the (0,0) beam appears as a sharp single peak whereas the $(\bar{1},0)$ beam is split into two beams of equal intensity

spots of equal intensity when a minimum of the diffraction function of the step array coincides with the maximum of the diffraction function of the terrace structure. With increasing voltage, i.e., decreasing wavelength, the maxima of the diffraction function due to the step array contract towards the direction of the specularly reflected beam with regard to the vicinal surface normal (ϕ_s in Fig. 2.7). The maxima of the terrace diffraction function move towards the specular beam with regard to the terrace normal. This implies that the two respective sets of maxima move across each other. In directions which correspond to maxima of the terrace diffraction function the diffraction function of the step array will exhibit, with changing wavelength, alternately maxima and minima.

The condition for this to happen does not depend on the terrace width and step orientation. Coincidence of maxima of both diffraction functions is achieved when scattered waves from atoms of neighboring terraces are in phase. Coincidence of a minimum of the diffraction function of the step array with a maximum of the terrace diffraction function occurs when atoms of neighboring terraces scatter in antiphase. Let $\underline{a}$ and $\underline{b}$ be the lattice vectors of the terrace structure and $\underline{a}^*$, $\underline{b}^*$ the corresponding reciprocal vectors ($\underline{a}\underline{b}^* = \underline{a}^*\underline{b} = 0$, $\underline{a}^*\underline{a} = \underline{b}^*\underline{b} = 2\pi$). Neighboring atoms of adjacent terraces may be connected by a vector $\underline{s}$ with component d (step height) normal to the terrace and lateral component $x\underline{a} + y\underline{b}$ expressed by fractions x and y of the terrace lattice vectors. For a given incident wave vector $\underline{k}_o$ the condition for maxima of the terrace diffraction function, is according to simple kinematic diffraction theory, given by

$$(\underline{k}_{h,k} - \underline{k}_o) \quad = h\underline{a}^* + k\underline{b}^*. \tag{2.5}$$

$$|\underline{k}_{h,k}| = |\underline{k}_o|. \tag{2.6}$$

Equation (2.5) describes the relation between the wave vector component of the incident and scattered beams parallel to the terrace plane and (2.6) gives the condition for elastic scattering. The conditions for in-phase scattering from adjacent terraces may be expressed by

$$\underline{s}(\underline{k}_{hk} - \underline{k}_o) = 2n\pi. \tag{2.7}$$

and for antiphase scattering by

$$\underline{s}(\underline{k}_{hk} - \underline{k}_o) = (2n + 1)\pi. \tag{2.8}$$

with n integer.

From (2.5-8) the distinct incident wave vectors $\underline{k}_o$ can be determined for which inphase or antiphase scattering occurs for a given reflex order (h,k).

Suppose $\underline{k}_o$ forms the angle θ with the terrace normal and the projection of $\underline{k}_o$ into the terrace plane has azimuthal angles ϕ_a and ϕ_b with respect to the reciprocal lattice vectors $\underline{a}^*$ and $\underline{b}^*$, respectively. Then $|\underline{k}_o| = 2\pi \sqrt{U/150}$ for which inphase scattering occurs is given by

$$|\underline{k}_o| = \frac{\pi}{d} \frac{(n-hx-ky)^2 + (h\underline{a}^*+k\underline{b}^*)^2 \, d^2/4\pi^2}{(n-hx-ky)\cos\theta - (h|a^*|\cos\phi_a+k|b^*|\cos\phi_b) \, d \, \sin\theta} \qquad (2.9)$$

with n integer. Half integer values for n yield antiphase scattering. Backscattering (only observed with LEED) implies

$$(n-hx-ky)^2-2(n-hx-ky)(h|a^*|\cos\phi_a+k|b^*|\cos\phi_b) \, d\tan\theta > (h\underline{a}^*+k\underline{b}^*) \, \frac{d^2}{4\pi^2} . \qquad (2.10)$$

Relation (2.9) is equivalent to an Ewald construction within the three-dimensional reciprocal lattice of Fig. 2.6 with the lattice vector $\underline{c}^* = (2\pi/d)\underline{e}$. $\underline{e}$ denotes a unit vector in the direction of the terrace normal. The condition for constructive interference of scattered waves from neighboring terraces is equivalent to the fulfillment of the third Laue condition.

Let us now return to the diffraction from vicinal surfaces with periodic step structures of step height d and terrace width Λ. Half integer values of n in (2.9) lead to the condition for $\underline{k}_o$ for which the diffraction function of the step array has minima in the direction of the diffracted beams $\underline{k}_{h,k}$, i.e., for which the terrace diffraction function exhibits maxima. In these directions no intensity will be observed. However, close to these directions maxima of the diffraction function of the step array occur which lead to satellite intensities as schematically sketched in Fig. 2.7. It appears as if the diffraction spots due to the maximum of the terrace diffraction function are split into two spots of equal intensity.

The splitting separation depends on the terrace width Λ: the smaller the terrace width the larger the splitting. We may describe the folding of the two diffraction functions by the Ewald construction of Fig. 2.8. The reciprocal lattice of the terrace structure is identical to that shown in Fig. 2.6. Superimposed is the reciprocal lattice of the periodic step array being inclined to the former by the angle α between the vicinal surface and the terrace plane. The dashed rods are separated by $(2\pi/d)\sin\alpha = 2\pi/|A|$ with $|A| = \Lambda/\cos\alpha$ the periodicity distance on the vicinal surface and Λ the terrace width. According to (2.9) these rods have to intersect the vertical rods of the former lattice in the points marked by the open circles characterizing the third Laue condition. If the Ewald sphere intersects the reciprocal lattices in these points, sharp single diffraction spots result because all surface atoms scatter in phase in the respective directions. For direc-

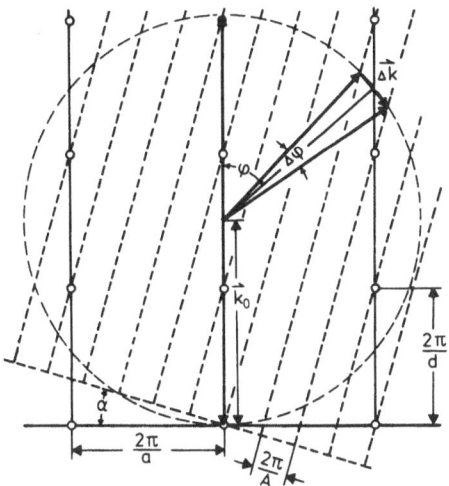

Fig. 2.8. Ewald construction applied to stepped surfaces for normal incidence of the primary beam to the terrace plane. The reciprocal lattice of the step array is inclined by the angle α towards the reciprocal lattice of the terrace structure. The angular separation of the split (1,0) beam is indicated by the wave vector difference Δk. The (0,0) beam appears as a sharp single beam because the Ewald sphere intersects the reciprocal lattice rods in a reciprocal lattice point characterizing the third Laue condition

tions crossing the rods midway between those points (as shown in Fig. 2.8 for the scattering angle ɸ), adjacent terraces scatter in antiphase. However, constructive interference of scattered waves from equivalent atom rows of the step structure occurs in two neighboring directions, separated by the angle Δɸ. From Fig. 2.8 (normal incidence) this angular separation can be determined as

$$\Delta\phi = \frac{2\pi}{d|\underline{k}_0|} \frac{\sin\alpha}{\cos(\phi-\alpha)} = \frac{\lambda}{\Lambda\cos\phi + d \sin \phi} \qquad (2.11)$$

and yields the angular distance of the split diffraction spots as function of wavelength λ, terrace width Λ, and scattering angle ɸ. A more general relationship for the three-dimensional case has been given by BESOCKE and WAGNER /29,133/.

The direction of the vector Δk describing the difference between the two wave vectors of the split beams is perpendicular to the step edge direction. The observed splitting direction on the LEED screen allows, therefore, the determination of the edge orientation.

In summarizing the results of this section we can conclude that LEED allows one to determine the following parameters of periodic step structures: 1) step height by recording distinct beam voltages at which, for a given diffraction order, single or split spots appear; 2) terrace width by measuring the separation of the split spots; and 3) orientation of the step edges by determining the directions of the spot splitting.

Figure 2.9 shows a LEED pattern of the W(750) surface taken at 110 V beam voltage. The (02) diffraction orders show split beams, whereas the (11) and (20) reflexes appear as single spots. The splitting direction indicates that the step edges run parallel to the [001] direction. The evaluation of the splitting separation yields a terrace width of 12.6 ± 0.1 Å /29/. The step height was determined

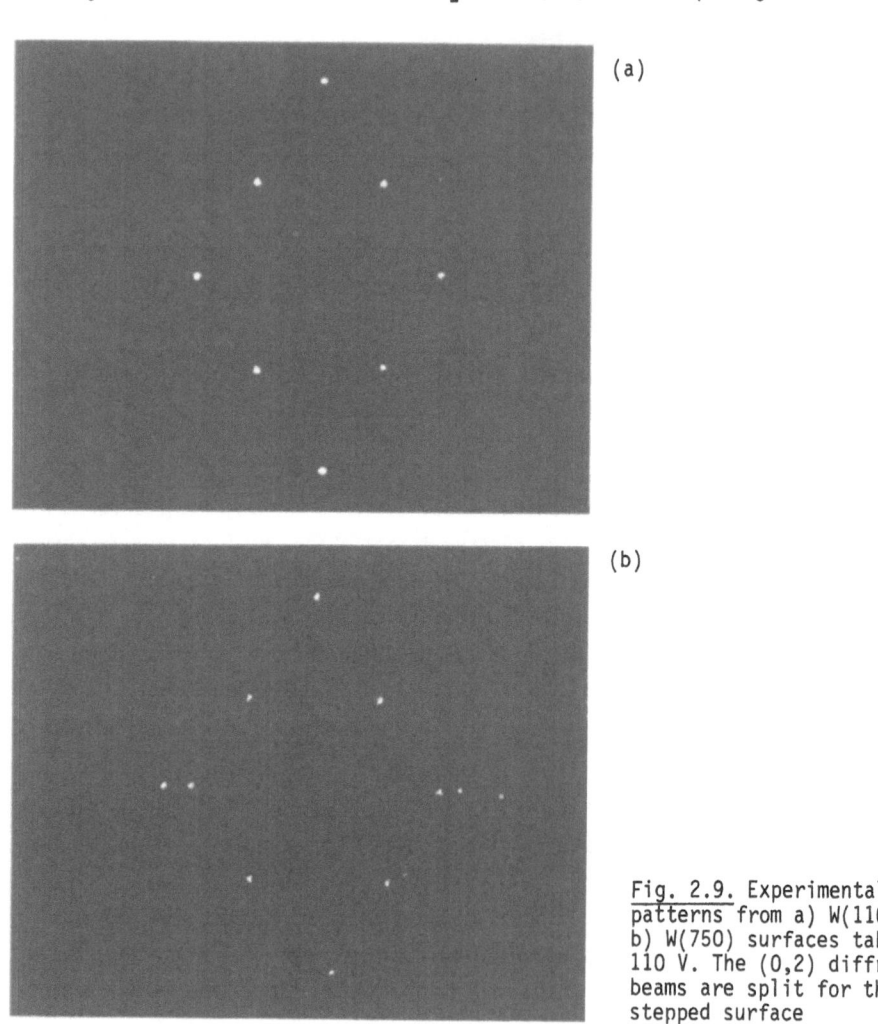

(a)

(b)

Fig. 2.9. Experimental LEED patterns from a) W(110) and b) W(750) surfaces taken at 110 V. The (0,2) diffraction beams are split for the stepped surface

according to (2.9) for $\theta = 0$. Figure 2.10 shows a plot of $|k_o|(n-hx-ky)$ versus $(n-hx-ky)^2$ for the (1,1) reflex at which sharp single and double spots of equal intensity appear. (x,y) may in this case be chosen as (1/2,0) and $\underline{a}$, $\underline{b}$, $\underline{a}^*$, $\underline{b}^*$ are given by the known structure of the W(110) plane. The step height resulting

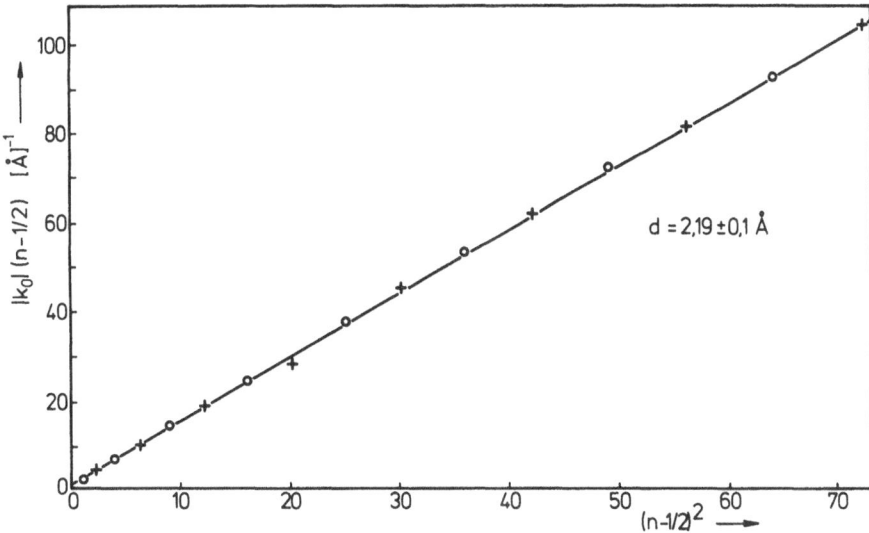

Fig. 2.10. Plot for determining the step height according to (2.9). Alternating appearance of single spots (+) and double spots (o) for the (11) reflex from the stepped W(650) surface (according to /29/)

from the slope of the plot was determined to 2.19 Å which is very close to the interplanar separation of the W(110) planes of 2.23 Å.

So far we have implicitly assumed a perfectly periodic step structure, i.e., all terraces have the same width Λ. This assumption may not be justified for real vicinal surfaces and the question arises what influence a distribution of terrace widths and/or step heights exerts on the observed LEED patterns. HOUSTON and PARK /36,37/ showed theoretically that LEED is rather insensitive to such distributions as far as the sharpness of diffraction spots is concerned. The step height and terrace width derived from LEED observations [(2.11) for terrace width] are virtually equal to the average values of step height and terrace width as determined from the corresponding distribution functions. Within experimental errors this conclusion holds for step height and terrace width distributions having half width values less than about half the corresponding average values /29/. We shall return to the question of nonequally spaced steps on macroscopically flat surfaces in Section 3.2.2.

HENZLER /25/ showed that irregular step structures produced by ion bombardment of GaAs surfaces give rise to LEED patterns with sharp or diffuse integral order spots depending on the electron beam voltage. In this case antiphase scattering of neighboring terraces results in the appearance of diffuse spots because no distinct terrace width or step orientation prevails which is the prerequisite for the occurrence of beam splitting. At certain beam voltages the various diffraction spots sharpen up due to a distinct step height between terraces leading to inphase scattering of all terraces with random width and orientation.

2.2.4 Electron Microscopy

Step structures on alkali halide surfaces have been made visible by transmission electron microscopy (TEM) using suitable decoration and replica techniques. The microscopic surface structure of cleavage planes /13,14,38/ and vicinal surfaces /15/ was obtained by gold deposition and subsequent carbon coating. The gold atoms coalesce into clusters preferentially at steps so that after dissolution of the alkali halide crystal the carbon replica carries a direct reproduction of the step distribution and can be studied by TEM. Step structures evolving during growth /13/ and evaporation processes /14/ and step motion associated with these processes have been clearly demonstrated and evaluated in view of theoretical model considerations /4/. Figure 2.11 reproduces /13/ the step structure around a screw dislocation

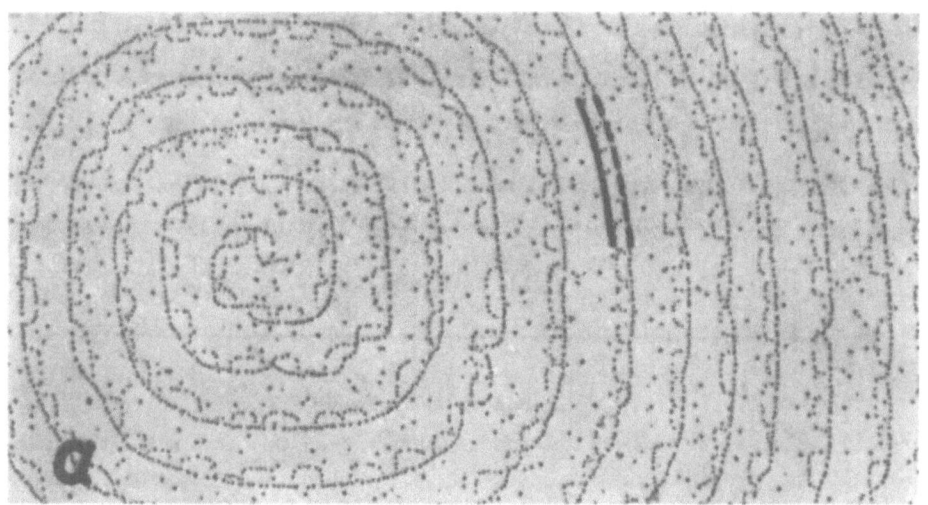

Fig. 2.11. Step structure around a screw dislocation on a cleaved (100) NaCl surface. Steps (dashed line) were first obtained by evaporation of the sample surface at 400 °C. Subsequently a supersaturation was applied at 330 °C, giving rise to regrowth of the spirals (solid line) (according to /13/)

which develops during evaporation and indicates further step motion after a subsequent growth process (distance between dashed and solid line). Figure 2.12 shows the step structure of a cleaved (100) NaCl surface in the steady state during evaporation /14/. The steps are directly visible and exhibit extremely uniform separations. According to BASSET /16/ the step height mostly equals one lattice plane distance. BETHGE et al. /13,14/ also observed steps of double height. REICHELT /15/

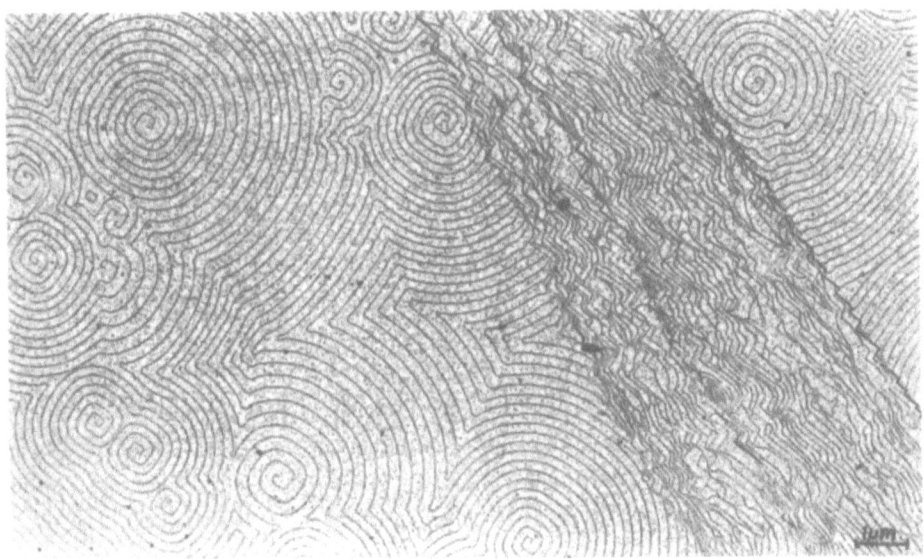

Fig. 2.12. Surface step structure on an evaporated NaCl crystal in the steady state (385 °C) showing well ordered step spirals around several screw dislocations (according to BETHGE, KELLER /14/)

investigated vicinal NaCl surfaces and observed regular step structures after annealing the samples which had been prepared by polishing. The surface normal of the vicinals was determined by the Laue technique yielding the average inclination angle towards the (100) terraces. From the separation of the steps ranging between 20-120 Å for the various vicinals, step heights of one and two lattice spacings were obtained. The occurrence of double step heights prevails for vicinals of larger inclination angles (for orientations (h 1 0) with h ≤ 8).

3. Properties of Clean Stepped Surfaces

3.1 Structural Properties

As described in the preceding sections there are essentially three experimental methods available to obtain structural information on stepped surfaces. Step separations on vicinal alkali halide surfaces can be measured directly by the decoration and replica technique combined with transmission electron microscopy. On these surfaces single and double step heights are present /15/. The steps do not, therefore, form an ideal, periodic array but rather adjust their separation on the average to comply with the inclination of the macroscopic surface. This fact may be due partly to the condition under which the vicinal surface structure develops during the necessary annealing treatment after the polishing preparation. On the other hand, the step structures evolving in the vicinity of dislocations on cleaved alkali halide surfaces show nearly ideal features with single step heights and extremely regular step spacings as depicted in Fig. 2.12.

So far no direct reproductions of step structures on vicinal metal and semiconductor surfaces are available because, for these materials, the decoration and replica method has not appeared to be applicable. Information obtained by LEED is in some sense indirect because only the Fourier transform of the surface structure can be evaluated. This means that values of step spacings and step heights are average values over distances of the order of the coherence length of the electron beam ($\sim$100 Å). These values can, however, be determined with high precision. As an example we consider results obtained from three tungsten (110) vicinals the (750), (650) and (40 37 1) surfaces. Table 3.1 /29/ summarizes values of the terrace width evaluated from the splitting of various diffraction beams. The standard deviation of the mean values is less than 1 % for the highly stepped (750) surface (large splitting) and increases to about 6 % for the surface with the largest terraces (small splitting). These values are not simply integral multiples of the separation between atom rows on the terrace. As mentioned earlier, LEED averages over an eventual terrace width distribution. If two possible terrace widths are necessary to yield the inclination of the macroscopic surface, the respective average value will be measured by LEED. The results given in Table 3.1 are, within experimental errors, compatible with the orientation of the macroscopic surface as determined by Laue back reflection.

Table 3.2 summarizes the results on the step height determination as obtained for various diffraction orders (h,k) by using (2.9). The standard deviation amounts to 0.5 to 1%. For a step height of one interplanar spacing between W(110) planes, one would expect a value of 2.23 Å according to the W lattice constant of 3.16 Å. The values derived by LEED are slightly smaller and seem to decrease with decreasing terrace width. The slightly lower values cannot be explained by a uniform depression of the upper atomic layer because the step height in this case still equals the bulk layer spacing. In the case of Ge and Si, HENZLER and CLABES /39/ observed a more pronounced decrease in step height of up to 3 % depending on terrace width. This decrease was considered to be due to a lowering of atom positions in edge sites by 0.25 Å. Several experimental findings as well as bond strength arguments are consistent with this explanation. A first theoretical treatment on steric properties of stepped surfaces was reported by TSANG and FALICOV /40/.

Table 3.1. Calculated terrace widths of three stepped W surfaces evaluated from various LEED reflections

| Diffraction order | Terrace widths for various sample orientations (Å) | | |
| (hk) | (750) | (650) | (40 37 1) |
	[7(110)x(1$\bar{1}$0)]	[12(110)x(1$\bar{1}$0)]	[19(110)x(2$\bar{1}$0)]
(00)	12.7	23.2	53.7
(11)	12.5	24.3	54.4
(1$\bar{1}$)	12.7	24.8	49.8
(02)	12.2	22.7	48.4
(0$\bar{2}$)	12.8	22.0	-
(20)	12.6	22.5	-
Average terrace width	12.6 ± 0.1	23.3 ± 0.5	51.6 ± 1.5

Table 3.2. Step height d (in Å) obtained from various LEED reflections for three W(110) vicinals

| Reflection | (750) | (650) | (40 37 1) |
	[7(110)x(1$\bar{1}$0)]	[12(110)x(1$\bar{1}$0)]	[19(110)x(2$\bar{1}$1)]
(00)	2.18 ± 0.01	2.21 ± 0.02	2.22 ± 0.01
(11)	2.18 ± 0.01	2.19 ± 0.01	2.19 ± 0.01
(20)	2.21 ± 0.01	2.19 ± 0.02	2.20 ± 0.01
(02)	2.18 ± 0.02	2.19 ± 0.02	2.18 ± 0.01

Ordered step structures on metal surfaces usually exhibit monoatomic step heights. One exception was observed by BLAKEKY and SOMORJAI /41/ in the case of a Pt(110) vicinal showing double step heights. The (110) face of the fcc Pt lattice is the least densely packed among the low index planes. On vicinals of the more densely packed low index planes of various metals so far investigated, only monoatomic step heights have been observed by LEED. As mentioned above, a mixture of single and double step heights would yield a measured value which equals the respective average value.

LEED studies on semiconductors reveal, however, the presence of double step heights, OLSHANETSKY et al. /26/ investigated (111) and (100) vicinals of germanium. They found that the vicinals with (111) terraces inclined towards the [211] direction exhibit double step heights of 6.28 Å which is by 4 % less than twice the value of the separation between the double layers of the diamond structure. These edges are not observed on cleaved surfaces /39/. For vicinals inclined towards the [211] direction, however, a step height of 3.14 Å was observed which is by 4 % less than the double layer spacing. The reduction of 4 % is consistent with the findings of HENZLER and CLABES /39/. Also on Ge(100) vicinals with edges parallel to the [011] direction, a step height of somewhat less (6 %) than two interplanar distances has been found

Similar investigations on ZnO (zinc blende structure) /27/ revealed the minimum step height on cleavage planes parallel to the C axis and steps of double height on cleaved polar planes perpendicular to the C axis. In this case, no reduction in step height relative to the bulk spacings was observed within the experimental errors.

Vicinal surfaces inclined towards a low index plane in such a way that the resulting edges are not parallel to a low index direction or, in other terms, contain a large number of kinks, form often ordered step structures, too. Those surfaces have been studied by LEED in the case of W /29/, Pt /41/, and UO_2 /42/. The observed splitting directions indicate that the edges run, on the average, parallel to the directions expected from the macroscopic surface orientation. ELLIS /42/ concluded by comparing LEED patterns with laser simulation patterns that, within a given edge, positions of kinks are randomly and not periodically spaced. Vicinals with high kink concentrations show a marked tendency to form hill and valley structures, i.e., they are composed of microfacets of neighboring orientations. This feature, intimately related to the thermal stability of the surface structure of vicinals, will be treated in the next section.

3.2 Thermal Stability of Step Structures

3.2.1 Experimental Observations

As described in a previous section, stepped surfaces may be prepared either by
cleavage or by standard metallographic procedures. In the first case the overall
surface orientation is that of the low index cleavage plane. Regions of various
step densities and step orientations are more or less randomly distributed over
the so prepared surface. Temperature treatments of cleaved surfaces will tend to
establish the low index plane all over the surface because this is the state of
lowest surface energy. If, on the other hand, a macroscopically flat vicinal surface prepared by usual metallographic techniques is given an annealing process,
its topographical microstructure will also try to comply with the energetically
most favorable state. This can be a regular step structure or a hill and valley
structure formed by facet planes of lower Miller indices. In the following section
we summarize the conditions which have to be met for the development of either of
the two topographies.

LEED observations regarding the thermal stability of step structures have been
reported for various metals and semiconductors. HENZLER and CLABES /39/ observed
that step structures on cleaved surfaces of the semiconductors Ge and Si disappeared
upon heat treatments at about half the melting temperature. On portions of the sample surfaces where the average inclination angle to the low index plane amounts to
several degrees, wide (111) terraces and steep planes of unknown orientations are
observed. Whether the step structures are indeed unstable should, however, be better explored on vicinals of large extensions. OLSHANETSKY et al. /26/ investigated
Ge vicinals close to the (111), (100), and (110) orientations. They found that the
(111) and (100) vicinals form ordered step structures et least up to 850 $^{\circ}$C which
is already close to the melting point of 937 $^{\circ}$C. If the step structures are de-
stroyed by argon ion bombardment they can be restored by temperature anneals above
600 and 350 $^{\circ}$C for the (111) and (100) vicinals, respectively. These observations
show that stepped Ge(111) and (100) vicinals are stable and yield the state of
lowest surface free energy. Ambivalent structural features are noticed for the
Ge(110) vicinals. In the lower temperature regime the vicinals facet, whereas at
higher temperatures the regular step structure forms the stable configuration.
These transitions are reversible. The transition temperature depends on which zone
the vicinal belongs to. These findings show that the structural features are equilibrium properties and do not depend on kinetic restrictions.

LEED patterns from cleaved Ge(111) and (100) faces exhibit different superstructures when raising the temperature over certain values. The superstructures correspond to atom rearrangements in the uppermost layer. For larger terrace widths on
stepped Ge surfaces the superstructures are also visible /26,27/. The transition

temperature between the different structures depends on step density /27/. The superstructure spots sometimes split into doublets like the integral order spots /26/ which shows that long-range order of the atom rearrangement is preserved across steps.

ELLIS /42/ found that regular step structures on UO_2 (111) vicinals are stable up to about 700 °C. Between 700 and 900 °C the ledges rotate and finally decompose irreversibly into microfacets. Thus the structural equilibrium state seems to be the hill and valley structure.

On metal vicinals, regular step structures are predominantly observed and are stable up to temperatures close to the melting point. This statement holds especially for vicinals of close-packed low index planes with edge directions parallel to low index directions. Vicinals of Cu(100) /23/, Pt(100) and (111) /8/, W(110) /29/, Pd(111) /32/, Ni(100) /31/, Ni(111) /30/, Au(111) /33/, Ir(111) /35/, and Re(0001) /24/ have been characterized by LEED and found to form stable regular step structures. The development of stepped surface regions on polycrystalline gold surfaces has been observed /43/ after argon ion bombardment and thermal treatment at 600 K. The most comprehensive study on the stability of vicinal surface structures was reported by BLAKELY and SOMORJAI /41/ for the case of Pt. Vicinals of the close-packed (111) plane showed regular step structures with monoatomic steps over the entire temperature range and irrespective of edge direction. Step structures on (001) vicinals occur for edges running parallel to the close-packed [110] direction, but faceting is observed for (001) vicinals belonging to the [100] zone. (011) vicinals belonging to the same zone show step structures of double height steps, whereas (001) vicinals of the [0$\bar{1}$1] zone exhibit facet structures. Vicinals of the (112) plane in the direction of the (111) plane show faceting only above 950 °C. The variety of structural features is summarized in Fig. 3.1 as taken from BLAKELY and SOMORJAI /41/.

The structural properties may be drastically changed by impurity adsorption at elevated temperatures. Usually the tendency for facet formation is enhanced by this process. BLAKELY and SOMORJAI /41/ also investigated the influence of oxygen and carbon adsorption on the structural properties of the Pt vicinals. The effect of the different impurities may be quite different depending on surface orientation. There are also indications /43/ that impurities can stabilize regular step structures.

From the experimental observations reported above one may conclude that the regular step structure on vicinal surfaces is one possible equilibrium state of lowest free surface energy. The other state conforms with a hill and valley structure set up by lower index facet planes. The differences in free surface energy of the two possible states may be quite small and may change sign with temperature, as indicated by the reversible surface structures of the Ge(110) vicinals /26/. If the vicinals show irreversible faceting at higher temperatures, one may regard

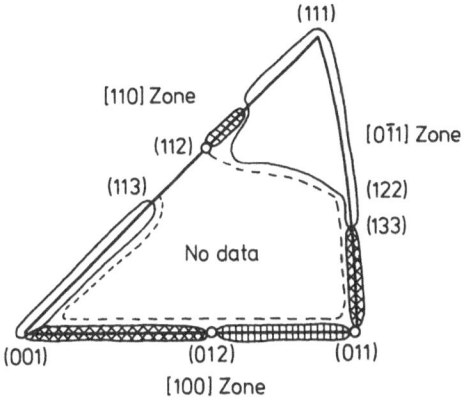

(111)

[110] Zone

(112)

(113)

[0̄11] Zone

(122)

(133)

No data

(001)   (012)   (011)

[100] Zone

⬭ monatomic height step
⬯ multiple height step
⬮ facet above 950 °C
⬮ facet

Fig. 3.1. Structural features of Pt surfaces indicated by various symbols within the stereographic triangle (according /41/)

this as the state of lowest energy in the entire temperature range and one might attribute the initial missing of well developed facets to the rather low mobility of surface atoms not allowing for formation of the equilibrium structure.

3.2.2 Theoretical Considerations

Let us first review some theoretical considerations regarding the stability of clean crystal surfaces. The problem is closely related to the orientation depend-ence of the surface free energy γ. The well-known WULFF construction /44/ estab-lishes the equilibrium shape of a crystal for which the total surface free energy attains a minimum. Suppose the dependence of γ on orientation is plotted radially as function of the surface normal $\underline{n}$ (γ plot, Fig. 3.2). The equilibrium crystal shape is then obtained by drawing lines (planes in the three-dimensional case) perpendicular to the radius vector at each point of the γ plot. Then the area (volume in three dimensions) which can be reached from the origin without cros-sing any of the lines (planes) has a shape similar to the equilibrium shape of the crystal. If the γ plot exhibits strong cusps in certain directions, the equi-librium crystal shape will be a polyhedron formed by planes whose normals coin-

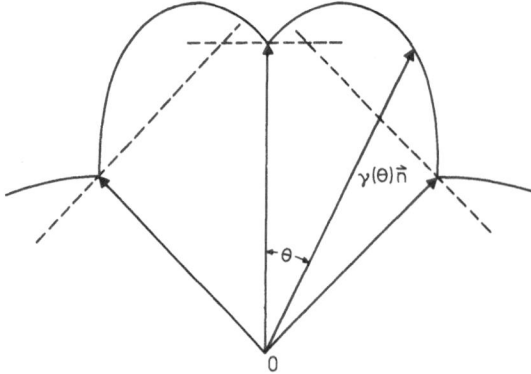

Fig. 3.2. γ plot showing the surface free energy as function of crystallographic orientation described by the surface normal n. Dashed lines form a polyhedron whose shape is similar to the equilibrium crystal shape

cide with the cusp directions. These directions correspond to orientations of low Miller index. All planes that are not part of the equilibrium shape are unstable.

Stepped surfaces are vicinals to low index planes and their γ values are described by portions of the γ plot in the vicinity of the low index cusp directions. Suppose a macroscopically flat crystal surface is prepared displaying an orientation close to a low index direction. The question arises whether the equilibrium surface structure of this vicinal plane will be formed by terraces of the low index plane and steps of monoatomic height with a density corresponding to the nominal surface direction or whether a hill and valley structure formed by low index planes (facets) will lead to a lower value of the surface free energy per unit area. HERRING /5/ showed that a potential reduction of surface energy by the formation of facets can be derived from the γ plot. In Fig. 3.3 let OA be the direction of the vicinal surface in question and OB_1, OB_2, OB_3 (the latter not being in the plane of the drawing) be three directions having positive projections on OA. The planes perpendicular to these three directions could form the facets on the macroscopic surface normal to OA. The surface free energy of such a faceted surface per unit area of the macroscopic surface would be

$$\gamma_s = \gamma_1 f_1 + \gamma_2 f_2 + \gamma_3 f_3. \tag{3.1}$$

γ_1, γ_2, γ_3 designate the surface energies of the three facet planes and f_1, f_2, f_3 are the areas of the planes per unit projected area in the plane of the macroscopic surface. If n_1, n_2, and n_3 denote unit vectors in the direction of the three

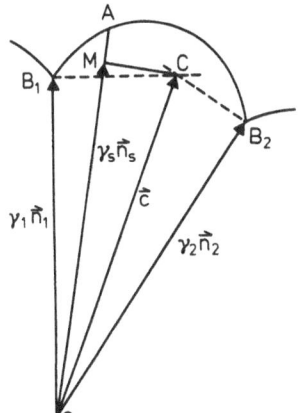

Fig. 3.3. Section of γ plot illustrating reduction in surface energy for direction n_s by a hill and valley structure formed by facets of orientation n_1 and n_2

facet planes, the unit vector $\underline{n}$ in the direction $\underline{OA}$ is given by

$$\underline{n} = f_1\underline{n}_1 + f_2\underline{n}_2 + f_3\underline{n}_3. \tag{3.2}$$

HERRING introduced the reciprocal vectors $\underline{n}^*_\nu$ given by

$$\underline{n}^*_1\underline{n}_1 = \underline{n}^*_2\underline{n}_2 = \underline{n}^*_3\underline{n}_3 = 1, \quad \underline{n}^*_\nu\underline{n}_\mu = 0 \quad (\nu \neq \mu). \tag{3.3}$$

and showed that γ_s can be expressed by

$$\gamma_s = \underline{c}\,\underline{n} \tag{3.4}$$

with $\underline{c}$ given by

$$\underline{c} = \gamma_1\underline{n}^*_1 + \gamma_2\underline{n}^*_2 + \gamma_3\underline{n}^*_3. \tag{3.5}$$

$\underline{c}$ represents the vector joining the origin with the point C obtained as the intersection of the three facet planes drawn normal to $\underline{OB}_1$, $\underline{OB}_2$ $\underline{OB}_3$, respectively. This is easily seen by the relations

$$\underline{c}\underline{n}_\nu = \gamma_\nu \quad (\nu = 1,2,3). \tag{3.6}$$

$|\gamma_s|$ is, according to (3.4), given by the projection of $\underline{c}$ on the normal $\underline{n}$ of the macroscopic plane and is represented by the length OM in Fig. 3.3. If, therefore,

M lies inside the γ plot, the macroscopic vicinal surface may lower its surface free energy by forming facets with normals $\underline{OB_1}$, $\underline{OB_2}$, $\underline{OB_3}$, respectively. If M lies on the γ plot, a facet structure is energetically not more favorable than the terrace-step structure of the ideal vicinal surface.

A sufficient condition for the γ dependence in the vicinity of a cusp direction leading to stable vicinal surfaces may be derived from (3.4). It is easily verified that the orientation dependence of γ_s is represented by a sphere through the origin and the three facet orientations $\underline{OB_1}$, $\underline{OB_2}$, $\underline{OB_3}$, respectively. If, therefore, the surface free energies of the vicinal planes lie on or within this sphere, the vicinal surface structures will be stable with regard to faceting. It is interesting to note that within an attractive pairwise interaction model /5/ the theoretical γ plot consists of cusps connected by smooth spherical portions. Such a model assumes that the crystal energy can be calculated by summing up the attractive pair interaction energies between all pairs of atoms. The forces between the pairs have to be of finite range and the atomic spacings should be the same within the lattice and in the surface region. As HERRING /5/ pointed out, the surface energy calculated within the framework of such a model can be represented by an equation of the form of (3.4).

Spherical portions in the vicinity of a cusp direction can also be phenomenologically rationalized in terms of a step structure for which an additional step energy β per unit step length is introduced /5/. Neglecting any interaction energy between steps, Fig. 3.4a illustrates that the orientation dependence γ(θ) can be expressed by

$$\gamma(\theta) = \cos\theta\left(\gamma_0 + \frac{\beta}{d}\,\tan\theta\right). \tag{3.7}$$

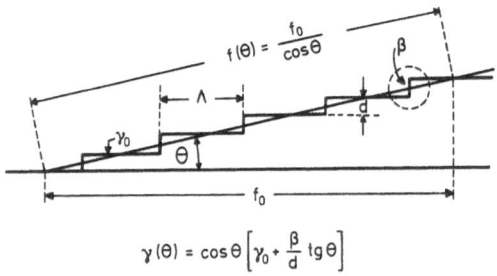

$$\gamma(\theta) = \cos\theta\left[\gamma_0 + \frac{\beta}{d}\,tg\theta\right]$$

Fig. 3.4a. Orientation dependence of surface energy γ(θ) by assuming a regular step structure and an additional step energy β per unit step length

with γ_0 the surface energy of the low index cusp orientation, d the step height, and θ the angle between the vicinal and the low index surface direction, respectively. $\tan\theta/d$ equals the number of steps per projected area in the low index surface. Figure 3.4b shows that (3.7) yields a spherical θ dependence with radius

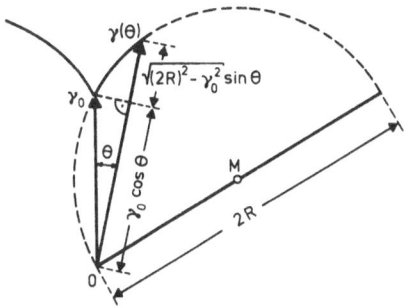

<u>Fig. 3.4b.</u> Spherical γ plot in the vicinity of a low index orientation derived from (3.7). The radius R of the sphere is related to the step energy β introduced in Fig. 3.4a

$R = \frac{1}{2}\sqrt{\gamma_0^2 + (\beta/d)^2}$. The larger the step energy β, the larger is R and the steeper the cusp at γ_0. The step structure of the vicinal surface will be stable with regard to faceting if the spherical portion of the γ plot described by (3.7) lies on or within the sphere given by (3.6). This condition may either be fulfilled or not depending on the vicinal surface orientation and the material as described in the previous experimental section. For instance, edges with a high concentration of kinks will exhibit higher β values than close-packed edges, and the corresponding step structures will accordingly show different tendencies towards faceting.

So far we have treated steps as isolated entities with no mutual interactions and not taking into account configurational changes of the edge structure at elevated temperatures, i.e., edge roughening which causes a lowering of the step free energy. Phenomenologically, a mutual step interaction energy can be accounted for by adding in (3.7) higher power terms in $\tan\theta$ /45/

$$\gamma(\theta) = \cos\theta[\gamma_0 + (\beta/d)\tan\theta + \alpha_2\tan^2\theta + \alpha_3\tan^3\theta + ...]. \tag{3.8}$$

Potential interactions between steps characterized by the coefficients α_ν were considered by BLAKELY and SCHWOEBEL /46/ who suggested longer ranged elastic interactions between steps and by WYNBLATT /47/ who chose a pairwise interaction model for the crystal atoms which were allowed to relax their positions in the outermost layers. The step interaction is expected to be repulsive on a vicinal surface. Another type of repulsive step interaction results from the change of configurational entropy /48/ connected with a single step edge with temperature. By the production of kinks the step edge attains a certain waviness which gives rise to a repulsive force between adjacent edges because decreasing the step spacing restricts the density of kinks. GRUBER and MULLINS /48/ applied geometrical and statistical mechanical principles to the terrace-ledge-kink model and calculated numerically the kink entropy contribution to the ledge free energy for various edge structures on an fcc surface. The free energy decreases with increasing temperature and tends to zero at a critical temperature which is discussed below. GRUBER and MULLINS /48/ thus obtained the temperature dependence of the term linear in $\tan\theta$ in (3.8) but derived for the coefficient β/d also a θ dependence. In view of the series expansion in (3.8), this θ dependence formally gives rise to nonvanishing α_ν values which in turn characterize mutual step interactions. According to the calculated results, these interactions have a repulsive nature and imply that closely spaced edges tend to exhibit lower kink concentrations than more widely spaced edges.

Repulsive step interactions would tend to keep the steps equally spaced on vicinal surfaces, i.e., the steps are ordered in a periodic array. In view of the above mentioned insensitivity of LEED to deviations from strictly periodic step structures /36,37/, some further indications will be given which support the assumption of equally spaced steps. The step arrays on alkali halide surfaces observed by electron microscopic (Figs. 2.11 and 2.12) represent periodic structures. The terrace widths are even much larger than those determined by LEED on vicinal metal surfaces. SCHWOEBEL and SHIPSEY /49/ and SCHWOEBEL /50/ treated theoretically the kinetics of step motion during condensation and evaporation. Assuming directionally dependent capture probabilities of terrace adatoms to steps, the results obtained show that either coalescence of steps or stabilization of step spacings can occur, depending on the inequality of the capture probabilities. If, therefore, faceting as a consequence of step coalescence is not observed, the stationary step structure should exhibit a distinct terrace width. The capillarity driven shape changes of metal surfaces also support the assumption of equal step spacings /51/. In this case, surface diffusion (the predominent kinetic process) provides for the mass transport and causes the flattening of, for example, an initial sinusoidal surface profile. The variation of the chemical potential represents the driving force for the mass transport and depends on the curvature of the macroscopic surface. An inhomogeneous distribution of steps is equivalent to a local curvature and should therefore anneal out, i.e., transform into a stable configuration formed by an ordered array of monoatomic steps with constant separation. The alternative to surface flattening would be bunching of steps of surface faceting (hill and valley structure) which results from the pecularity of the γ plot and may in special cases offset the effect of local curvature.

The temperature range where the edge free energy approaches zero was investigated by LEAMY and GILMER /51/. These authors applied the Monte Carlo simulation technique to vicinal surfaces of a simple cubic lattice. In this case, not only the production of kinks within steps present on the vicinal surface at low temperatures is simulated but also the production of adatoms, vacancies, and additional edges on the terraces at higher temperatures. By incorporating the effect of surface roughening on the terraces, the energy and the free energy of an isolated edge tend to zero at a certain critical temperature. Close to this temperature the structure of the vicinal surface exhibits steps of multiple height units and the initial step structure is no longer distinguishable. This finding confirms the original suggestion of BURTON and CABRERA /52/ that surfaces undergo a roughening transition at a critical temperature which causes singularities in the structural and thermodynamic properties of the surface. However, the experimental verification of a roughening transition below the melting point of metals has, to the knowledge of the writer, not yet been established. Besides a potential faceting of vicinal surfaces, the predominant process to occur on clean stepped surfaces consists of an increasing production of kinks within the initially straight edges with increasing temperature. The kink concentration as a function of temperature and edge orientation was derived by BURTON, CABRERA, and FRANK /4/.

Another important process can have a dramatic effect on the step structure of vicinal surfaces, i.e., impurity adsorption which can either occur from the gas phase or may be due to surface segregation of bulk impurities. GJOSTEIN /53/ treated the influence of impurity adsorption on the surface free energy, i.e., on the γ plot. Two major cases can be considered: 1) the impurity atoms are more tightly bound in edge positions which will therefore be occupied first, or 2) the terrace sites will be the more favorable positions. Figure 3.5 illustrates the influence of impurity adsorption on the angular dependence of the surface free energy in the neighborhood of a cusp direction. In case of preferential terrace adsorption, the surface free energy of the terraces will be lowered and the cusp will be more pronounced. If the adsorption takes place mainly at edge positions, the surface free energy in the cusp direction is not changed, but the angular dependence shows deviations with regard to the slope and the torque of the γ plot. As GJOSTEIN /54/ showed, faceting may result from both cases considered as caused by changes in the respective γ plots. The reverse can happen, too. ISA et al. /43/ found that by calcium segregation on polycrystalline Au samples the formation of periodic step

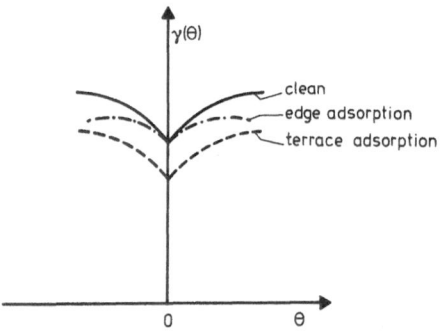

<u>Fig. 3.5.</u> Schematic representation of the effect of impurity adsorption on the γ plot for vicinal orientations (according to /53/)

structures became obviously more favorable as indicated by the characteristic LEED pattern.

Another effect of impurity adsorption may simply consist of production of different step heights. ELLIS and SCHWOEBEL /22/ observed that annealing a stepped UO_2 crystal surface in vacuum caused steps of double height, while oxygen exposure led to steps of minimum height. One should therefore bear in mind that the adsorption of foreign atoms or molecules may change the step structure and thus can cause different adsorption properties.

3.3 Electronic Properties

3.3.1 Work Function

The electronic work function depends strongly on the structural arrangement of atoms in the outermost surface layers of solids. Close-packed surfaces exhibit higher work function values than open surface structures. The work function of various low index planes of the same metal may differ by up to 20 %. For instance, the close-packed W(110) plane has a work function of 5.2 eV /55/ whereas the open structure of the (111) plane yields a value of 4.4 eV /56/.

The physical reason for these differences is the surface dipole contribution to the work function which depends on the electron charge distribution in the surface region. This distribution is markedly affected by the surface structure. SMOLU-CHOWSKI /57/ correlated structure and charge distribution and showed that close--packed atom arrangements give rise to dipole contributions which cause higher work function values than those connected with open surface structures. Figure 3.6 illustrates qualitatively this reasoning. The positive ion charges are smeared out over the appropriate Wigner-Seitz cells and the mobile electron charges adjust their local distribution to minimize the sum of kinetic and potential energy. The relaxation of negative charges can be thought of as a "smoothing" effect resulting in a dipole moment per unit surface area which tends to lower the energy barrier for electron escape and hence to lower the work function. Work function calculations based on a self-consistent treatment of the inhomogeneous electron gas /58/ confirm this argument.

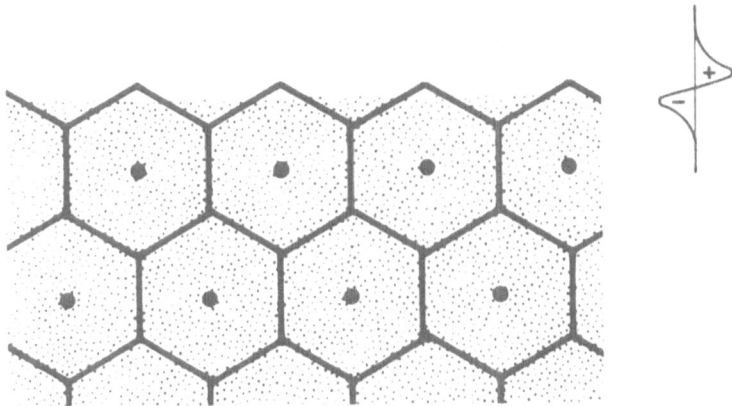

Fig. 3.6. Qualitative description of the "smoothing effect" of the electron density distribution within the surface region. The positive ion charges are homogeneously distributed over the marked Wigner-Seitz cells of the respective crystal structure. The mobile electron charges relax parallel to the surface, giving rise to a surface dipole which decreases the work function (according to /57/)

Steps on surfaces represent structural features where charge redistributions may occur and give rise to additional surface dipoles. KRAHL-URBAN et al. /55/ investigated the work function of various vicinals of the W(110) plane by thermionic emission measurements. Surfaces characterized by the stereographic triangle shown in Fig. 3.7 have been studied. The edge structures corresponding to the three different edge orientations are indicated on the respective zone axis. Figure 3.8 reproduces the work function dependence on step density for the [001] zone, i.e., the edges are parallel to the [001] direction. The work function decreases linearly with step density. This result can be rationalized in terms of dipole moments

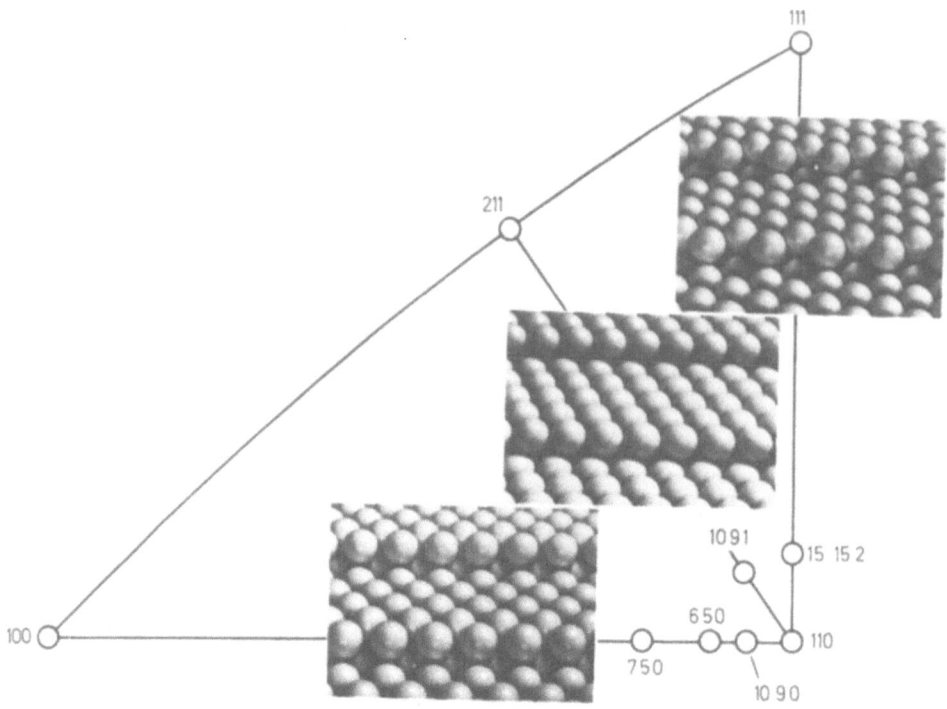

Fig. 3.7. Stereographic triangle depicting surface orientations and step structures of W(110) vicinals investigated by work function measurements /55/

associated with the edge atoms or, correspondingly, with a dipole moment per unit
edge length. According to the Helmholtz equation

$$\Delta\phi = 300 \cdot 10^{-18} \ 4\pi n\mu \tag{3.9}$$

the work function change $\Delta\phi$ in eV is related to the number n per cm^2 of dipole moments μ given in Debye units (D). Evaluation of the slope in Fig. 3.8 yields
$\mu = 0.37 \pm 0.03$ D per edge atom of a step in the [001] direction.

    For the edge structures exhibited by the vicinals belonging to the other zones
(Fig. 3.7), slightly different dipole moments are obtained. Table 3.3 summarizes
the results. The close-packed edge parallel to the [1Ī1] direction shows the
smallest dipole moment, whereas the more open edge structures give rise to larger
values. This finding agrees with SMOLUCHOWSKI's /57/ consideration on the "smoothing" effect which causes larger work function reductions for more open structures.

    BESOCKE and WAGNER /59/ investigated work function changes during the deposition
of W on a W(110) plane. The observed work function reductions could be explained
by considering growing islands whose edges give rise to additional dipole moments.

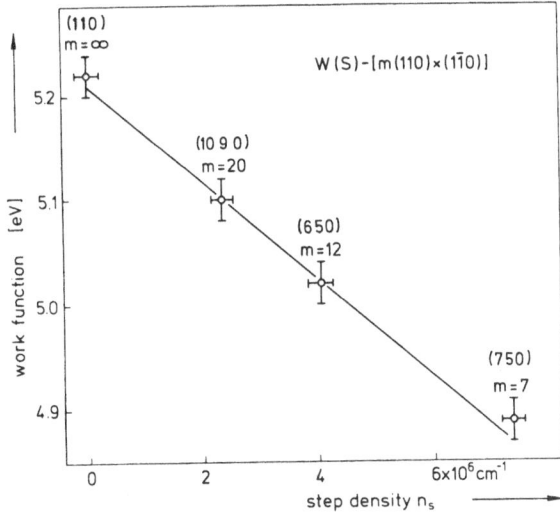

Fig. 3.8. Work function of W(110) vicinals belonging to the [001] zone as function of step density /55/

An average dipole moment per lattice constant of 0.3 D was evaluated, in excellent
agreement with the results of KRAHL-URBAN et al. /55/. In the same study a dipole
moment of about 1 D was obtained for a single adsorbed W atom on the W(110) plane.
This value is about three times larger than the values for edge atoms which seems
quite reasonable in view of the larger smoothing effect expected for this configuration.

Table 3.3. Dipole moments per lattice constant for edges of various orientations
on W(110) vicinals (KRAHL-URBAN et al. /55/)

Step direction parallel to	Step orientation	Dipole moment (D)
[$\bar{1}$11]	(101)	0.29 $\pm$ 0.04
[$\bar{1}$10]	(112)	0.34 $\pm$ 0.04
[001]	($1\bar{1}0$)	0.37 $\pm$ 0.03

Work function reductions have also been observed during evaporation of W on stepped W surfaces with (110) terraces but different step densities /134/. For room temperature adsorption, the total work function changes observed decreased with increasing step density. The results are easily interpreted by considering that with increasing step density a growing fraction of deposited W atoms are able to diffuse to nearby edges. Only the fraction of W atoms forming clusters and hence additional edges causes the observed work function reduction. Under the deposition condition used (pulsed deposition of 0.02 monolayer at a time), virtually no work function change could be observed on a stepped surface with an average terrace width of 12 Å. This result is in accord with the measured W adatom diffusivity on a W(110) plane obtained by field ion microscopy /60,61/.

KÖRNER /62/ deduced the work function of W(100) vicinals using a rather indirect statistical method for evaluating measurements on recrystallized W foils. He inferred a parabolic dependence of work function on step density. A linear dependence, however, could have likewise been inserted in the evaluation.

The work function measurements on the W(110) vicinals by KRAHL-URBAN et al. /55/ were performed in a wide temperature range (2200-2800 K) and allowed the determination of the temperature coefficient of the work function. The absolute values of the temperature coefficient, being negative for all investigated surfaces, decrease with step density by as much as a factor of two. Edge structures with smaller dipole moments show larger absolute values of the temperature coefficient.

BESOCKE et al. /34/ investigated the work function dependence on step density and edge structure for Pt and Au. In this case cylindrically shaped surfaces were prepared with the close-packed (111) plane in the central part and stepped portions of increasing step density on either side. The stepped portions showed the characteristic beam splitting in LEED observations. The step edges were parallel to the close-packed [1$\bar{1}$0] direction. Due to the trigonal symmetry of the (111) plane, the edge structures are different for the two cylindrical sides. On the two sides the step edges are formed by the (100) and (11$\bar{1}$) planes, respectively. In SOMORJAI's notation /8/ the stepped surfaces can be characterized by S[m(111)x(100)] and S[m(111)x(11$\bar{1}$)] with m the number of atom rows on the terraces decreasing from the central part ot the sample edge. Work function changes relative to the value of the central (111) plane were measured as a function of sample position, i.e., as a function of step density. Figure 3.9 reveals again linear work function reductions with step density for both edge structures and both metals. The structure corresponding to the more open (100) plane shows the larger dipole moments for both metals. The dipole moments for Pt are more than a factor of two larger than the corresponding Au values.

As far as the dipole moment dependence on edge structure is concerned, SMOLU-CHOWSKI's "smoothing effect" provides a qualitative theoretical explanation. A quantitative theoretical description is extremely difficult because of the intri-

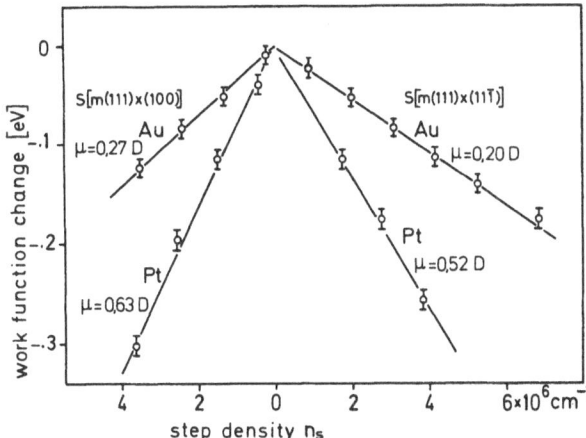

Fig. 3.9. Work function variation versus density for Pt and Au surfaces with (111) terraces and steps parallel to the close-packed [1$\bar{1}$0] direction. Different edge structures give rise to different dipole moments μ (BESOCKE et al. /34/)

cate problems involved in treating self-consistently the inhomogeneous electron gas near surface steps. The same problems certainly arise in rationalizing the differences in dipole moments for various metals. Again, rather qualitative considerations may help to unravel the underlying physical reason. KESMODEL and FALICOV /63/ estimated the electron charge transfer from d orbitals of edge atoms and found this to be proportional to the density of d states near the Fermi level. The latter quantity depends on the chemical nature and takes much higher values for Pt than for Au and W. Consequently the dipole moment associated with a Pt edge atom should be larger than that for a W or Au edge atom in an equivalent structural configuration. This qualitative prediction is confirmed by the experimental results so far obtained.

HUISER and VAN LAAR /64/ found work function differences between stepped and smooth portions for cleaved GaAs surfaces. For p-type material the stepped portions exhibited a lower and for n-type material a higher work function than the corresponding smooth parts. For this semiconducting material, the authors explained the work function changes by corresponding band bendings due to surface states created by steps. Evidence for surface states associated with steps will be given in the following section.

3.3.2 Surface States

Work function variations due to steps give evidence for local redistributions of
electron charges. A change in the energy distribution of electron states in the
surface region might be expected as well. It is well known that additional electron states which are energetically different from bulk states (surface states)
are localized in the uppermost surface layers or that the local density of states
is much enhanced in this region (surface resonances). The occurrence and the energy distribution of surface states depend markedly on the structural arrangement of
the top layers /65/. Steps on surfaces might therefore change the local density of
states or even give rise to additional surface states. First experimental evidence
for additional surface states due to steps was reported by HENZLER and CLABES /39/.
They found a monotonic dependence of the surface photovoltage on step density on
cleaved Si(111) surfaces. The photovoltage equals the work function change due to
illumination. A subsequent investigation /66/ revealed that this effect is caused
primarily by an enhanced carrier recombination via surface states associated with
steps. In this respect, different step orientations give rise to different photovoltage changes which might be due to a different distribution of surface states.

   ROWE et al. /67/ observed step-dependent variations in the ultraviolet photoemission spectra from cleaved Si(111) surfaces. Structures of the photoelectron
emission spectra are related to the density distribution of occupied electron
states. Besides surface states associated wiht dangling bonds and back bonds on
the flat Si surface, an additional surface state due to steps has been found 0.4 eV
higher than the main surface state close to the upper valence band edge. Further
photoemission effects due to steps were reported /68,69/ for badly cleaved GaAs(110)
surfaces.

   The experimental work has stimulated theoretical investigations on the electronic structure of stepped surfaces. RAJAN and FALICOV /70/ applied a tight-binding
approximation to the Si(111) surface with up and down steps in the close-packed
[1$\bar{1}$0] direction. They found a sizeable rearrangement of the sp^3 orbitals at atoms
with dangling bonds, giving rise to an increased charge in direction of the dangling bonds. The Fermi level is lowered by 0.3 eV with respect to the unreconstructed Si(111) surface which has also been observed experimentally /67/. New
step states appear both above and below the normal Si(111) surface states. Step
states associated with double dangling bonds which are characteristic for atoms
located in steps of single height are found below the other surface and step states.
The charge transfer resulting from the rearrangement of the sp^3 orbitals of step
atoms should yield a sizeable dipole moment which tends to increase the work function. An experimental verification of this effect has not yet been demonstrated.
Step-dependent surface states on Si(111) were also found theoretically by SCHLÜTER
et al. /71/ who employed the self-consistent pseudopotential method. Additional

step-induced states showed up 0.4 eV below the Fermi level and could be attributed to step atoms with two dangling bonds. States due to step atoms with one dangling bond nearly coincide with the (111) terrace dangling bond states and are only partially occupied

Experimental evidence for surface states or resonances on metal surfaces has been obtained in only a few cases, i.e., for W /72,73/ by measuring the energy distribution in field emission or by angle resolved photoemission. With the latter technique HEIMANN et al. /74/ showed that surface states exist on the (111) faces of the noble metals Cu, Ag, and Au. Differences in the photoemission spectra /75/ from the reconstructed and unreconstructed Pt(100) surface revealed features which could be interpreted as surface resonances due to the structure of the unreconstructed surface. Additional states due to surface steps on metals have not yet been verified experimentally.

Theoretical investigations of metal surfaces showed, however, that the local density of states depends quite markedly on the structural arrangement of surface atoms. DESJONQUERES and CYROT-LACKMANN /76/ applied a tight-binding approximation to various low index planes of Ni(fcc) and Fe(bcc). They found that especially the less dense-packed planes exhibit strong features in the local density of states which are related to the atomic levels of the metal. In a subsequent investigation the same authors /77/ studied the stepped Pt surfaces Pt(S)[6(111)x(001)] and Pt(S)[9(111)x(01$\bar{1}$)]. We reproduce (Fig. 3.10) from this study the local density of states n(E) for a position within a large (111) terrace (dashed curve) and for the upper edge of the (001) step (solid curve). There is a striking feature around 0.2 Ry below the Fermi level for the stepped surface which is related to one of the 5d atomic levels which is especially affected by the reduced number of nearest neighbors.

Various cluster models have been treated to obtain the electronic structure of atomic clusters. For a limited number of atoms the eigenvalue problem is tractable and the local density of states for different atom positions may be obtained. Depending on the cluster geometry edge and terrace positions can be approximated and their influence on the electronic structure simulated. JENNINGS et al. /78/ used the multiple-scattering formalism and applied to various 3d transition metal clusters. Further investigations undertaken by cluster methods are cited in /78/.

Besides the so far reviewed electronic states at stepped surfaces, SALZBERG and GONCALVES DA SILVA /79/ investigated the spectral density of one-magnon states and demonstrated its dependence on atom position on a stepped surface with six unequivalent sites.

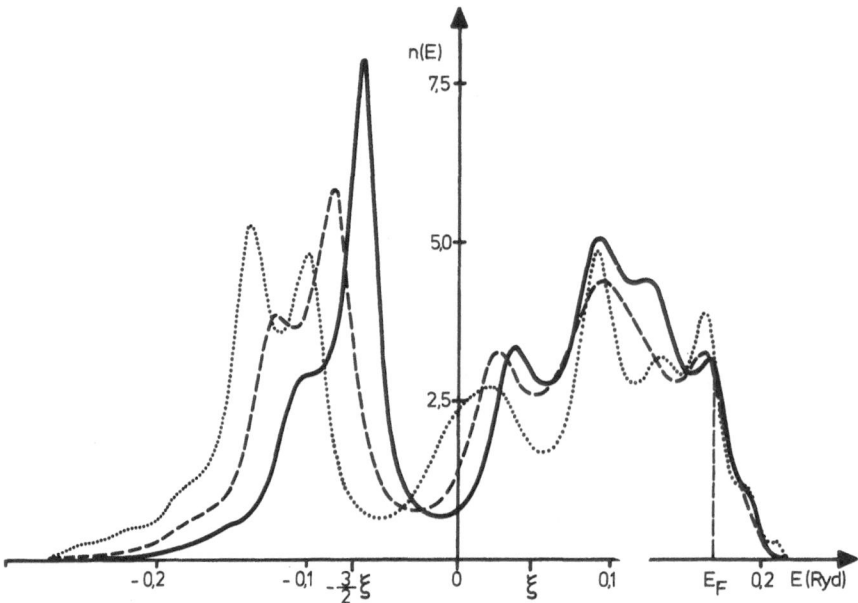

Fig. 3.10. Local density of states for Pt bulk (.......), Pt(111)(_____), and edge row of Pt(S)[6(111)x(001)] (————). Theoretical results by DESJONQUERES and CYROT-LACKMANN /77/

4. Interaction of Atoms and Molecules with Stepped Surfaces

The major incentive to study the physical properties of stepped surfaces as described in the preceding sections arises undoubtedly from the endeavor to understand in more detail the modified chemical behavior of stepped surfaces with regard to the interaction with foreign atoms or molecules. The importance of step and kink sites in condensation and evaporation is well documented in the literature on crystal growth. Excellent review articles have been published on this subject /80,4,81/ which will, therefore, not be discussed further. The tremendous importance of adsorption-desorption processes and dissociation and association reactions of and between adsorbed species on solid surfaces is well recognized in the field of heterogeneous catalysis and corrosion. Manifold experimental evidence has proven for a long time that these processes depend in many cases on surface structure. The concept of "active sites" in heterogeneous catalysis tries to correlate catalytic activity with special structural properties of the catalyst surface. In this sense step and kink sites may be thought of as promoting or at least influencing catalytic activities. The availability of stepped surfaces with preset step densities and step orientations enables systematic investigations on the influence of these structural arrangements.

The following sections deal with the influence of step structures on adsorption kinetics, adsorption states, and catalytic reactions as well as other kinetic processes, such as surface segregation and diffusion. An increasing number of publications treating this subject has appeared in recent years. In connection with LEED observations, first adsorption studies on regularly stepped Cu surfaces were reported by PERDEREAU and RHEAD /82/. Oxygen and H_2S adsorption proceed at a much faster rate on the stepped surfaces as compared to the low index (100) face. SOMORJAI and co-workers extensively studied gas-interactions with stepped surfaces. Their first investigation /8/ in this respect treated H_2, O_2, and CO adsorption on the catalytically important metal platinum. Again, gas adsorption readily occurred on the stepped surfaces and at a much slower rate on the smooth low index planes.

Two publications seem to have especially stimulated world wide interest in research on stepped surfaces. BERNASEK et al. /83/ and BERNASEK and SOMORJAI /84/ reported H-D exchange reactions measured by a molecular beam technique and found the exchange yield to be several orders of magnitude higher on stepped Pt surfaces than on the smooth Pt(111) plane. IBACH et al. /85/ observed that the initial stikking coefficient of oxygen on cleaved silicon samples increases exponentially with step density by more than two orders of magnitude. Although subsequent investigations /86-89/ did not show that the influence of steps on these kinetic processes causes enhancement of several orders of magnitude, they nevertheless proved that steps give rise to appreciable promoting effects.

4.1 Adsorption Kinetics

Quantitative data on the differences of adsorption kinetics encountered on smooth and stepped surfaces can only be reliably determined if accurate exposure measurements are possible. Usual partial pressure readings lack this accuracy. Therefore, special emphasis should be given to those experiments in which identical exposure conditions are applied to the respective sample surfaces. PERDEREAU and RHEAD /82/ as well as BESCOCKE and BERGER /90/ used samples which exhibited macroscopic facets of desired orientations. HOPSTER et al. /91/ employed a sample surface of cylindrical shape with varying step density in one direction. Under these conditions, relative differences in adsorption kinetics between stepped and smooth surface areas can be studied quantitatively.

BESOCKE and BERGER /90/ investigated oxygen adsorption on stepped tungsten surfaces with (110) terraces. The sample employed displayed the smooth (110) face in the central part which was surrounded by four macroscopically flat regions inclined by 5^0 and 10^0, respectively. Each of two opposite flats exhibited the same edge orientation, i.e., edges parallel to the [001] and [1$\bar{1}$0] directions, respectively, but different terrace widths (12.5 and 25 Å). Room temperature adsorption of oxygen proceeded with remarkedly different rates on these five surfaces. Figure 4.1 reproduces the oxygen sticking coefficient (fraction of impinging molecules which become adsorbed) as a function of oxygen coverage. The sticking coefficient on the stepped faces turns out to be much higher than on the (110) plane and takes sizeable values even for coverages above 0.5 monolayer where the oxygen uptake is very small on the (110) plane under the experimental conditions used (O_2 partial pressure between 1×10^{-9} and 2×10^{-8} Torr). The initial sticking coefficient (for $\theta \to 0$) shows a linear dependence on step density as shown in Fig. 4.2. The vicinal surfaces with steps parallel to the [001] direction [step orientation (1$\bar{1}$0)] give rise to a larger initial oxygen uptake than the vicinals with steps parallel to the [$\bar{1}$10] direction [step orientation (112)]. These results may be complemented by the work of ENGEL et al. /92/ who studied oxygen adsorption on two W(110) vicinals with steps running parallel to the [111] direction which is the close-packed direction in a bcc crystal structure. In this case no appreciable differences in the oxygen adsorption kinetics was observed between the stepped and the flat surfaces. It can therefore be concluded that the step structure, or more precisely the step orientation, plays an important role in effecting the adsorption kinetics. It remains to be seen which physical property of the structural arrangement of step atoms provides for the observed differences.

In this respect the work of KING /93/ on the adsorption of nitrogen on various W planes including highly stepped surfaces of the [001] zone deserves special attention. Refining a model first put forward by ADAMS and GERMER /94/, KING and WELLS /95/ proposed that a necessary requirement for the dissociative adsorption of N_2 on W surfaces is the presence of neighboring [001] sites. Once dissociated, the N atoms can move on to sites of different geometry, e.g., those present on the (110) plane. The enhanced chemisorption of N_2 on stepped W(110) surfaces was thus attributed /93/ to the presence of [001] sites within the steps.

HOPSTER et al. /91/ studied the catalytic oxidation of CO on stepped Pt(111) surfaces using a sample surface with varying step density. In this section we shall focus our attention only on their results pertaining to the oxygen adsorption kinetics. The sticking probability increased exponentially with step density expressed by

$$S(n) = S_0 \, e^{8,9 \, n} , \qquad (4.1)$$

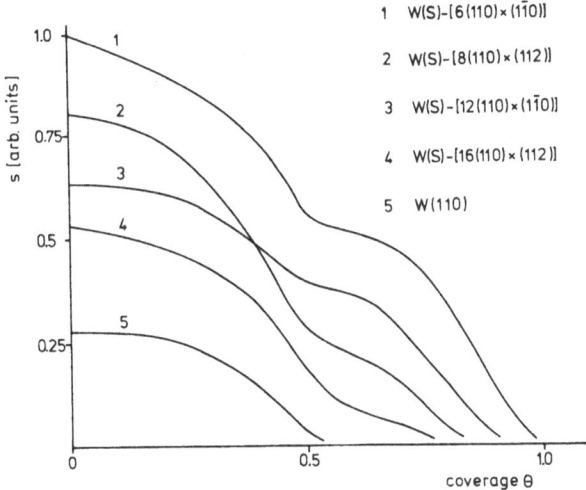

1 W(S)-[6(110)×(1̄10)]

2 W(S)-[8(110)×(112)]

3 W(S)-[12(110)×(1̄10)]

4 W(S)-[16(110)×(112)]

5 W(110)

Fig. 4.1. Sticking coefficient of oxygen on various stepped W(110) vicinals as a function of coverage (room temperature adsorption) (according to BESOCKE and BER-ger /90/)

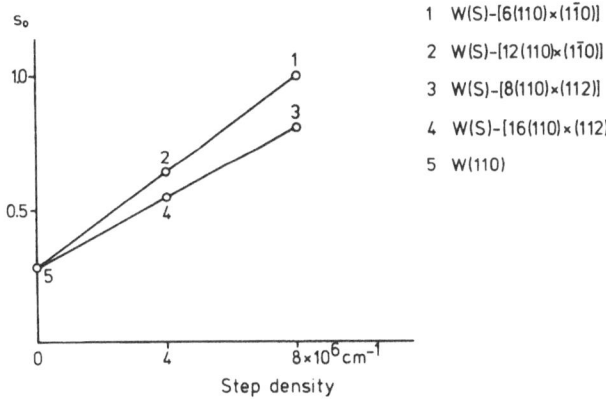

1 W(S)-[6(110)×(1̄10)]

2 W(S)-[12(110)×(1̄10)]

3 W(S)-[8(110)×(112)]

4 W(S)-[16(110)×(112)]

5 W(110)

Fig. 4.2. Initial sticking coefficient of oxygen as a function of step density for two edge orientations on W(110) vicinals (according to BESOCKE and BERGER /90/)

with S_o the sticking probability for the smooth Pt(111) surface and 1/n the number of atom rows per terrace. n varied between 0 < n < 0.1, i.e., in the region of highest step density the terraces contained on the average ten atom rows. Over this step density range the sticking probability increased by a factor of 2.4 according to (4.1).

In separate exposure runs, CHRISTMANN and ERTL /88/ investigated the hydrogen adsorption on Pt(111) and the stepped Pt(997) surface. Their result on the difference in sticking probability observed for the two surfaces is reproduced in Fig. 4.3a. The initial sticking probability on the stepped surface is roughly a

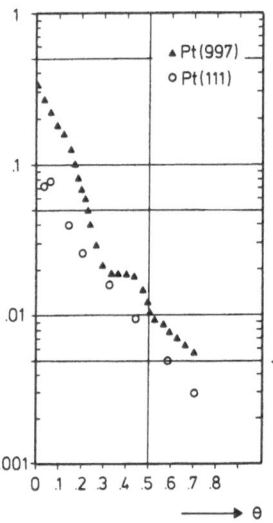

Fig. 4.3a. Sticking coefficient of hydrogen on Pt(111) (o) and Pt(997) (▲) as function of coverage for adsorption at 120 K (according to CHRISTMANN and ERTL /88/)

factor of four larger than on the flat (111) plane. The given absolute data are accurate only within ± 30 % due to the reasons given above and the coverage determination via flash desorption measurements. Nevertheless, the enhanced sticking probability on the stepped surface is clearly evident as well as the different functional dependence on hydrogen concentration. The authors /88/ pointed out that the two breaks in the curve for $\theta \sim 0.15$ and $\theta = 0.25$ for the stepped surface can be correlated with hydrogen adsorption on two different types of step sites. Further support for this interpretation is lent by the concomitant work

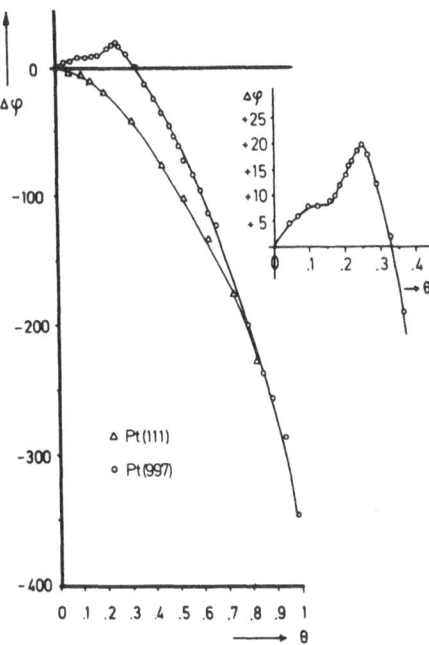

Fig. 4.3b. Work function change (mV) due to hydrogen adsorption at 130 K as function of coverage. The inset shows an enlarged portion of the plot for the stepped surface (according to CHRISTMANN and ERTL /88/)

function changes (Fig. 4.3b). Adsorption of hydrogen at step sites increases the work function, whereas adsorption on the Pt(111) plane leads to a monotonic decrease.

PIRUG et al. /96/ investigated the oxygen adsorption on a highly stepped Pt[4(100)x(111)] surface. The Pt(100) plane normally forms a reconstructed surface but can also be brought into a metastable unreconstructed (1x1) state. On the reconstructed (100) surface virtually no oxygen adsorption could be detected (detection limit for the sticking coefficient < 0.01). The unreconstructed as well as the stepped reconstructed surfaces give rise to oxygen adsorption. The sticking probability of the stepped reconstructed surface amounts to about 0.05. If the stepped surface exhibits unreconstructed terraces, the sticking coefficient takes a value of 0.17, whereas the corresponding value for the unreconstructed flat plane is 0.10.

The oxygen adsorption on the stepped Ni[4.5(111)x(11$\bar{1}$)] surface was studied by ERLEY /97/. Although oxygen is quite readily adsorbed on the smooth Ni(111) surface, the sticking coefficient on the stepped surface is larger by a factor of 1.4.

These examples may suffice to show that adsorption kinetics proceed at much faster rate on stepped surfaces, especially when the respective smooth low index planes exhibit small sticking coefficients. As illustrated by oxygen adsorption on stepped W surfaces /90/, the enhanced kinetic processes may strongly depend on the edge structure. This behavior raises the question of which specific physical properties associated with step edges give rise to the observed enhancement. The theoretical description and calculation of adsorption kinetics are at present not even tractable for well-characterized low index surfaces. One is therefore bound to consider more qualitative arguments for rationalizing step-enhanced adsorption kinetics. At first one may think of a step-promoted action localized at step sites due to unsaturated bonds or free valence orbitals of edge atoms. This hypothesis could be stressed by the different influences experienced by different edge structures /90,93/. In the case of dissociative adsorption (H_2, O_2, N_2) the interaction of the molecule orbitals of the gas molecule with the orbitals of the respective edge atoms may be stronger than the corresponding interaction with terrace atoms and may therefore lead to preferential molecular bond scission required for subsequent atomic adsorption. Effects of this kind will be discussed later in the section of catalytic processes.

A different view on the promoting action of steps was taken by IBACH /98/. He considered the dissociative adsorption into the chemisorbed state via an activation barrier from a weakly bound molecular precursor state. The height of the activation barrier depends on the electrostatic potential experienced by the precursor molecule. The potential in turn is changed (lowered) by the presence of dipole moments associated with the step edges as described in a previous section. Thus the activation barrier for chemisorption is decreased, which leads to enhanced chemisorption kinetics due to the presence of steps. Within this model IBACH /98/ and HOPSTER et al. /91/ could consistently correlate the variation of oxygen chemisorption kinetics on Pt, Si, and GaAs as function of step density and for the semiconductors also as function of doping and surface reconstruction. Further experiments and theoretical considerations are necessary to elucidate the physical parameters which are responsible for the altered adsorption kinetics of stepped surfaces.

4.2 Adsorption States

Although the most pronounced effects due to steps relate to variations in adsorption and reaction kinetics, there are a few examples reported in the literature which show that also the adsorption state, i.e., the binding energy of adsorbates depends on the presence of step sites. It is well known for various adsorbate-substrate systems that the binding energy depends on the crystallographic orien-

tation and thus on the surface structure. One may therefore expect an influence
of step sites on binding energy.

Isosteric heat and flash desorption measurements yield information on the adsorbate binding energy. The isosteric heat as a function of coverage is derived
from adsorption isobars or isotherms by making use of the Clausius-Clapeyron relation. In flash desorption experiments the substrate is covered with an initial
concentration of adsorbate and the temperature is subsequently raised linearly
with time or with another defined time dependence. The increase in partial pressure due to the desorbing species is recorded by a mass spectrometer. If the vacuum system is continuously pumped with high and constant pumping speed, the mass
spectrometer reading is directly proportional to the desorption rate. The temperature at which the desorption rate attains its maximum value relates to the desorption energy and yields (for associative desorption in conjunction with the
dissociation energy) the binding energy.

CONRAD et al. /32/ determined by isosteric heat measurements the adsorption
energy of hydrogen on the stepped Pd[9(111)x(111)] surface. For small coverages
the adsorption energy increased by about 3 Kcal/mole as compared to the smooth
(111) plane. In a subsequent paper, CHRISTMANN and ERTL /88/ found by flash desorption experiments that the desorption energy of hydrogen on the stepped
Pt[9(111)x(111)] increased with decreasing hydrogen coverage (θ_H < 0.2) from
9.4 Kcal/mole, the value for the (111) plane, up to 12 Kcal/mole. The increase
in desorption energy of about 3 Kcal/mole for both the stepped Pd and Pt surfaces was attributed to hydrogen adsorption at step sites. Taking into account
the H_2 dissociation energy of 103.2 Kcal/mole results in binding energies for a
chemisorbed hydrogen atom of 56.3 Kcal/mole for the flat surface and of 57.6
Kcal/mole for step sites. This small differences of 2.3 % already accounts for
the following observation on the work function change due to hydrogen adsorption
on the stepped Pt surface. CHRISTMANN and ERTL /88/ found that the maximum value
of the work function change as shown in Fig. 4.3b decreased with increasing temperature without changing the total amount of adsorbed hydrogen. Above room temperature virtually no work function change could be detected for hydrogen coverages θ_H < 0.3. The small difference in binding energy for step and terrace sites
is, at these temperatures, insufficient to cause exclusively the occupation of
step sites in this coverage range. The opposite effect of the two occupied sites
on the work function change causes the work function to stay virtually constant
up to θ_H = 0.3.

The experimentally determined small change in binding energy of 2-3 % is in
contrast with theoretical calculations of ZDANSKI and SROUBEK /99/ on hydrogen
adsorption on a stepped Ni(111) surface. These authors used the molecular orbital
scheme described by NEWNS /100/ which stresses the adatom coordination number and
neglects charge redistribution effects. A ratio of almost two resulted for the

binding energies at a step and a terrace site, respectively. The calculated difference is undoubtedly too large, and more refined theoretical treatments are necessary to achieve more realistic results.

HAGEN et al. /35/ reported differences in flash desorption spectra for oxygen
adsorption on Ir(111) and the stepped Ir[6(111)x(100)] surface as reproduced in
Fig. 4.4a,b. The peak temperatures of the flash desorption spectra for the Ir(111)
surface decrease with increasing exposure, i.e., increasing coverage which is indicative of a second-order desorption process. For small coverages the peak temperature is around 800 $^{\circ}$C. In the case of the stepped surface the maximum desorption rate occurs for small initial coverages at a higher temperature around 1300 $^{\circ}$C.
With increasing coverage a second desorption peak develops around 800 $^{\circ}$C. The high-
-temperature peak can be related to oxygen adsorption at step sites, whereas the
peak around 800 $^{\circ}$C corresponds to adsorption at terrace sites as for the smooth
Ir(111) plane. No evaluation in terms of desorption energies which would be required for the determination of the corresponding binding energies was given by
HAGEN et al. /35/. A very crude estimate based on the respective peak temperatures
yields a difference of roughly 10 % in the binding energies for step and terrace
sites.

Differences in binding energies on flat and stepped surfaces were also reported
by ISETT and BLAKELY /31/ for carbon on nickel. In this case the segregation equilibrium between a known carbon bulk impurity concentration and the carbon surface

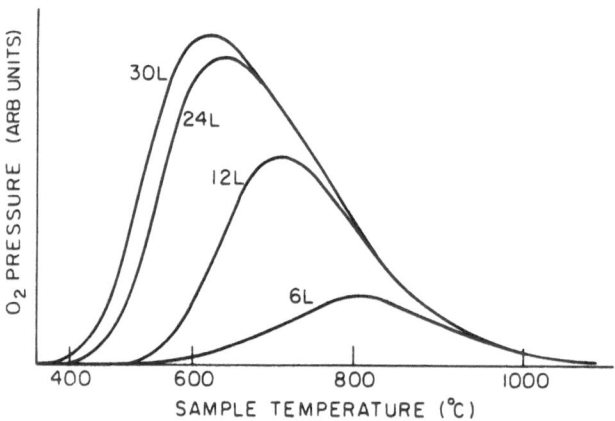

Fig. 4.4a. Thermal desorption spectra of oxygen adsorbed on Ir(111). The heating
rate was about 70 K/sec. Parameter is the oxygen exposure (1 L = 10^{-6} Torr sec)
(according to HAGEN et al. /35/)

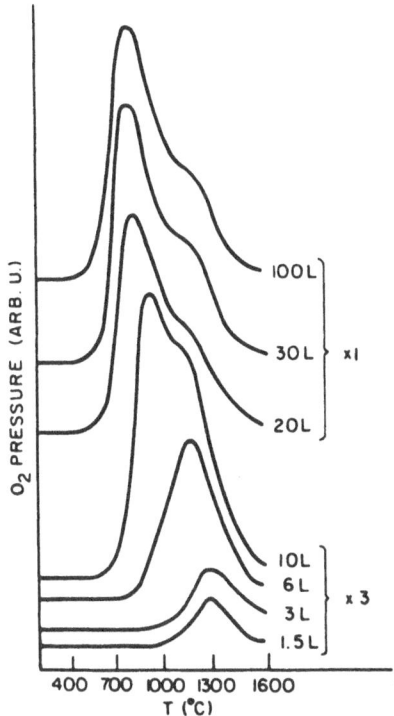

Fig. 4.4b. Thermal desorption spectra of oxygen adsorbed on Ir(S)-[6(111)x(100)].
Heating rate about 100 K/sec (according to HAGEN et al. /35/)

coverage θ was determined by AES as function of temperature. From a Langmuir plot,
i.e., the dependence of $\ln(\theta/1-\theta)$ on reciprocal temperature, binding energies were
deduced. For various coverage regimes characterized by distinct LEED patterns,
binding energies were obtained for the low index Ni(100) and the stepped Ni
Ni[7(100)x($\bar{1}$33)] surface. Table 4.1 gives the results as reproduced from ISETT
and BLAKELY /31/. The largest difference is obtained for small coverages (adsorption at step sites) and amounts to about 5 %. The step structure exerts a
much smaller influence on the binding energy for higher coverages, e.g. for gra-
phitic overlayers. This result is quite understandable by considering that the
interaction energy between the carbon atoms within a specific overlayer structure
is only slightly influenced by the presence of steps. The last column in Table 4.1
gives calculated binding energy values in terms of the CFSO-BEBO semiempirical
binding energy scheme /101,102/ and shows very good agreement with the experimental results.

Table 4.1. Binding energies of carbon atoms on Ni surfaces (according to ISETT and BLAKELY /31/).

Surface	Binding energy [eV]	Coverage	CFSO-[a] BEBO [eV]
C atom in solution in Ni	- 6.91		- 6.85
C atom on Ni(100) Phase I	- 7.38	Less than p(2x2)	-7.38
C atom on Ni(100) Phase II	- 7.11	Saturates at c(2x2)	
On Ni(S)[7(100)x($\bar{1}$33)]	- 7.23	Greater than 2/3 p(2x2)	- 7.41
In graphite	- 7.35		- 7.35
On Ni(S)[7(100)x($\bar{1}$33)]	- 7.36	1/3 p(2x2) $\leq \theta \leq$ 2/3 p(2x2)	- 7.50
Monolayer graphite on Ni(111)	- 7.40	Monolayer of graphite expitaxially bound to Ni(111)	- 7.40 ± 0.04
On Ni(S)[7(100)x($\bar{1}$33)]	- 7.71	$\theta \leq$ 1/3 p(2x2)	- 7.84

[a] Crystal field surface orbital: bond-energy-bond-order model calculations /101,102/.

4.3 Adsorbate Structures

Besides the influence on the interaction energy between adsorbed atoms or molecules, the terrace structure of stepped surfaces may limit the extension of ordered adsorbate structures. In some cases it has been observed that, nevertheless,
well-ordered adsorbate structures extend over several terraces. PERDEREAU and
RHEAD /82/ observed nitrogen and carbon overlayers extending over several terraces on stepped copper surfaces. CHESTERS et al. /103/ and ROBERTS and PRIT-
CHARD /104/ studied adsorption of the rare gases Ar and Kr on the stepped Ag(211)
and Cu(211) faces, respectively. Hexagonal overlayer structures formed irrespective of the stepped nature of the substrate surfaces. In many cases, however, the
terrace structure exerts a marked influence on the long-range periodicity of adsorbate overlayers. This is especially to be expected if the growing adsorbate
layer nucleates at step sites. In this respect three features imposed on the overlayer arrangement as elucidated by LEED observations are worth mentioning.

1) If the average terrace width is large enough to accommodate one or more unit
cells of the adsorbate superlattice (in the direction perpendicular to the steps),
registry between the superlattices on adjacent terraces may exist. In this case

the fractional order spots of the LEED pattern are split at distinct beam voltages like the integral order spots. As an example we refer to oxygen adsorption on stepped tungsten (110) surfaces with edges parallel to the [1Ī1] directions. This direction runs parallel to the most densely packed tungsten rows. On the flat W(110) plane, oxygen adsorption occurs in two equivalent domains of p(2x1) struc- ture which also form on the stepped surface. Figure 4.5 shows a structure model where the two domains are denoted I and II /92/. If on the regular step array oxy-

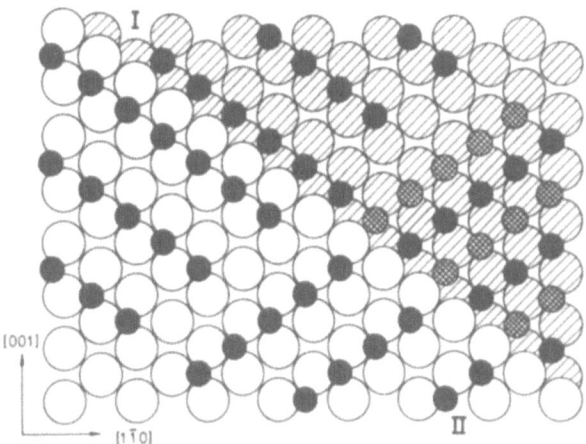

Fig. 4.5. Structure model of oxygen adsorption on a stepped W surface with (110) terraces and edges parallel to the [111] direction. Two (2x1) domains (type I and II) are possible. Type II domains may exhibit two different phase relations across steps as indicated by the different symbols on the lower terrace /92/. The cross- -hatched symbols have been added

gen adsorption starts at step sites, there will be a fixed phase relation between type I domains on adjacent terraces which gives rise to splitting of fractional order beams at distinct beam voltages. This behavior has indeed been observed by ENGEL et al. /92/. The type II domains on different terraces lack a fixed phase relation whether they nucleate at step sites or randomly on terraces. This is il- lustrated in Fig. 4.5 by the two possible and nonequivalent choices for placing the domain on the lower terrace relative to the arrangement on the upper terrace. Consequently, the fractional order beams due to type II domains are not split. For other step directions, both domains have no fixed phase relation across the steps

and hence no splitting of the respective fractional order spots is observed as
shown by BESOCKE et al. /105/.

2) As may be expected by inspection of Fig. 4.5, the growth of type I domains
is certainly preferred if nucleation starts at step sites. This effect was clearly demonstrated by ENGEL et al. /92/ by comparing the intensities of the fractional diffraction spots for the two domain structures as a function of oxygen coverage. Figure 4.6 shows normalized intensities for the (1/2, 1/2) diffraction spot
(due to domain type I) and the $(\bar{1}/2/\ 1/2)$ spot (due to domain type II) for the
W(S)-[10(110)x(011)] surface. (Note the different scales for the two domains!)
It can thus be concluded that the presence of steps leads to the preferential
growth of one of two possible domain structures which are equivalent on the respective flat low index plane.

3) Adsorbate structures which are well ordered on flat low index planes may
show a high degree of disorder or even lack long-range periodicity completely in
the presence of steps. BESOCKE et al. /105/ found only faint indications for the
development of the p(2x1) oxygen structure on highly stepped W(110) surfaces with
steps parallel to the [001] and $[\bar{1}10]$ directions. NIEUWENHUYS et al. /106/ observed
that carbon residues on stepped iridium (111) surfaces produced by adsorption of
various hydrocarbons form disordered structures, whereas ordered carbon layers develop on the flat Ir(111) plane. Hydrocarbon adsorption leads similarly to disordered carbonaceous layers on the stepped Pt[7(111)x(310)] surface with edges
containing a large number of kink sites /107/. The influence of steps on the ordering of adsorbate layers may in part be due to the enhanced adsorption kinetics
which renders the simultaneous ordering process more difficult. However, heat
treatments following the adsorption process did not lead to well-ordered structures /106,107/. This indicates that the respective step structures also influence
the interactions between the adsorbed species which are very important for the
development of ordered layer structures.

4.4 Catalytic Reactions

In this section we shall summarize the most important results on chemical reactions which adsorbed molecules undergo on stepped surfaces as compared to those
experienced on the corresponding flat surfaces. The investigations reported so
far deal with the bond breaking of simple diatomic molecules such as hydrogen
(isotope exchange reaction), carbon monoxide, nitric oxide; with oxidation reactions, for example, of CO; and with hydrogenation and dehydrogenation reactions
of various aliphatic and aromatic hydrocarbons.

Catalytic processes may often proceed via bond breaking reactions of some of
the constituents involved. In this context studies of dissociation reactions en-

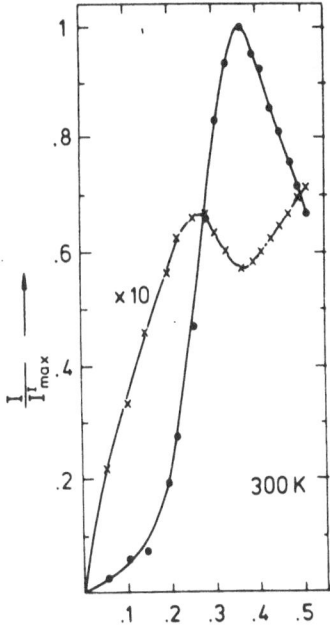

Fig. 4.6. Integral LEED intensity in the (1/2, 1/2) (●) and ($\bar{1}$/2,1/2) (x) diffraction spots as a function of oxygen coverage θ on the W(S)-[10($\bar{1}$10)x(011)] surface.
Adsorption temperature 300 K. Beam voltage 65 V (according to ENGEL et al. /91/)

hanced by steps deserve considerable attention. BERNASEK et al. /83,84/ investigated
the hydrogen-deuterium exchange reaction on flat and stepped platinum surfaces. A
molecular beam technique was employed where either a chopped beam of a H_2, D_2 mixture impinged on the Pt surface or a chopped D_2 molecular beam hit the surface with
an ambient H_2 background pressure present. Angle-resolved intensity measurements of
the backscattered species were carried out by a mass spectrometer. In this section
we shall focus our interest only on the detection of the reaction product HD. Whereas no HD molecules could be detected in this study from the smooth Pt(111) surface,
a HD intensity with an angular cosine distribution was observed for the stepped Pt
surfaces. The integrated reaction probability defined as total desorbed HD flux divided by D_2 flux was estimated to 10^{-1}. BERNASEK and SOMORJAI /84/ quoted for the
detection limit of the reaction probability a value of $<10^{-5}$ which results in the
conclusion that the stepped surfaces cause a reaction enhancement of more than four
orders of magnitude. According to the interpretation of BERNASEK and SOMORJAI, step
sites promote the adsorption and dissociation of H_2 or D_2 molecules. The chemisorbed
atomic hydrogen then reacts via a two-branch mechanism with D_2 molecules to form
the HD product. The drastic effect on the H-D reaction probability reported by
BERNASEK et al. /83,84/ could not be reproduced in similar studies /86-88/. Also

in the same laboratory GALE et al. /108/ found in a subsequent investigation an enhanced H-D reaction rate of less than an order of magnitude. LU and RYE /86/ and
CHRISTMANN and ERTL /88/ measured the steady-state H-D reaction rate at fixed H_2
and D_2 partial pressures as a function of sample temperature. In one case /86/ the
highly stepped Pt(211) and in the other case /88/ the stepped Pt[9(111)x(111)] surfaces were compared with the smooth Pt(111) plane. LU and RYE /86/ found the H-D
production rate to differ only by at most a factor of two and CHRISTMANN and ERTL
/88/ stated a factor of 10 ($\pm$ 50 %). The observed differences in production rate
could in both cases rather well be attributed to the differences in sticking probabilities of H_2 and D_2 on the two respective surfaces (see Sect. 4.1). WACHS and
MADIX /87/ reevaluated the experimental data of BERNASEK et al. /83,84/ and concluded that these data lead to results comparable with the results derived from the
stationary measurements /86/ by properly taking account of the nonstationary character of the molecular beam method. Nevertheless, the influence of steps on the
hydrogen-deuterium exchange reaction proves to be marked enough to cause considerable effects on the rates and selectivities of catalytic reactions involving adsorbed hydrogen /88/.

A recent investigation by GALE et al. /108/ showed that the H-D reaction rate
depends also on the reactant angle of incidence. An enhanced reaction probability
is observed if the reactant beam strikes the open side of the step structure. As
pointed out by the authors, this is perhaps the direction in which the available
bonding orbitals of the surface atoms are pointing. GARCIA /136/ rationalized the
enhanced H-D reaction rate on stepped surfaces in terms of a more efficient process
for impinging molecules to be scattered into bound states. Molecules in bound states
have larger life times and hence a higher probability to dissociate and to undergo
reactions with other adsorbed atoms.

Next we consider the influence of steps on the dissociation reaction of carbon
monoxide on various metal surfaces. CO may be adsorbed in either a molecular or a
dissociated state depending largely on the chemical nature of the substrate material. BRODEN et al. /109/ found that at a particular temperature a border line can
be drawn through the periodic table dividing the transition metals into two groups
for which predominantly molecular or dissociative adsorption prevails. Elements
close to the border line show an ambivalent behavior and a special influence of
the surface structure with respect to the CO adsorption state. It may therefore
be expected that steps markedly affect the adsorption state for this class of elements.

CO adsorption on flat and stepped surfaces of rhenium and nickel which belong
to this class have been studied by flash desorption measurements /110,30/. HOUS-
LEY et al. /110/ found on the basal Re(0001) plane two adsorption states, a lower
temperature α state (split into two substates) and a higher temperature β state.
The α state corresponds to molecular adsorption and exhibits for room temperature

adsorption a population which is by a factor of ten higher than that observed for the β state being attributed to dissociative adsorption. On the stepped Re(S)-[14(0001)x(10$\bar{1}$0)] surface, CO adsorption occurs first into the β state which in this case shows a population five times higher than that on the flat surface. The maximum population of the α state is virtually unaffected by surface structure. CO adsorption at elevated temperatures shows that the filling of the β state is a thermally activated process, leaving the question undecided whether the β state on the flat surface populates already at room temperature or during the flash desorption at elevated temperatures. In any case, the presence of steps markedly increases the dissociative CO adsorption on Re by lowering the respective activation barrier. CO desorption from the dissociative β state must proceed via a recombination of C and O atoms followed by desorption of the CO molecule. In this respect, it is worth mentioning that the peak temperature of the flash desorption spectrum corresponding to the β state is the same for both the flat and the stepped surface. This finding will be compared with observations on nickel surfaces as described next.

ERLEY and WAGNER /30/ studied CO adsorption on the flat Ni(111) and the stepped Ni(S)-[5(111)x(1$\bar{1}$0)] surface. The steps of the latter surface contain a high density of kink sites. Flash desorption spectra reveal that CO adsorbs on the Ni(111) only in one single molecular α state. However, electron beam interaction with adsorbed CO molecules causes dissociation and correspondingly leads to a second higher temperature peak ($β_1$ state) in the flash desorption spectrum. This peak is the only one which appears at exactly the same temperature if carbon and oxygen are separately brought onto the Ni(111) surface prior to the flash /30/. This observation unambiguously proves that the $β_1$ peak relates to the dissociative adsorption state.

Figure 4.7a,b shows the desorption spectra obtained from the flat Ni(111) surface. CO adsorption at room temperature on the stepped Ni[5(111)x(1$\bar{1}$0)] surface gives rise to two desorption peaks as shown in Fig. 4.7c. The lower temperature peak corresponds to the molecular adsorption state and appears at the same temperature as on the Ni(111) plane. The higher temperature $β_2$ peak occurs at an even higher temperature than obtained for the $β_1$ state on the flat surface and must also be attributed to a dissociative adsorption state. The $β_2$ state may be thought of as arising from the recombination of C and O atoms at step sites followed by the desorption of the CO molecule. A higher activation barrier for the C-O recombination might therefore be expected. This view is supported by the observation that upon dissociation on terrace sites, e.g., by electron beam interactions, the $β_1$ peak also appears as for the flat Ni(111) plane (Fig. 4.7d).

Whereas the same desorption state related to dissociative CO adsorption is observed on flat and stepped Re surfaces /110/, two peaks are found for Ni /30/ which can be correlated with desorption from terrace and step sites. One possible explanation for this difference might be the fact that associative desorption from

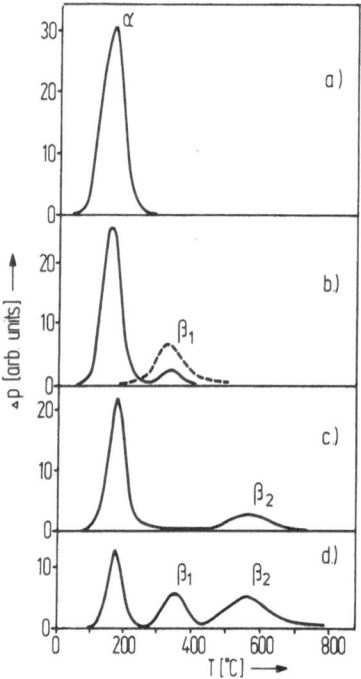

Fig. 4.7. CO flash desorption spectra from Ni surfaces (according to ERLEY and WAGNER /30/). (a) Ni(111), adsorption at room temperature. The α peak corresponds to molecular adsorption. (b) Ni(111), adsorption at room temperature followed by electron bombardment. Dashed curve: separate adsorption of C and O. The β_1 peak corresponds to dissociative adsorption. (c) Ni(S)-[5(111)x(1$\bar{1}$0)], adsorption at room temperature. (d) Ni(S)-[5(111)x(1$\bar{1}$0)], adsorption at room temperature followed by electron bombardment. The β_2 peak corresponds to dissociative adsorption at step sites

step sites on Re involves a rather high activation energy, so that only desorption from terrace sites occurs.

The associative desorption of CO or, in catalytic terms, the oxidation of carbon is rendered more difficult in the presence of steps as shown above for the Ni(111) surface. Catalytic reactions often involve several reaction steps which may be differently affected by steps. The adsorption process of the reactants may be enhanced and the reaction rate between the adsorbed species may be decreased, so that the net influence of steps on the catalytic process can be either positive or negative, depending on the reaction conditions. A rather illustrative example is the catalytic oxidation of carbon monoxide on stepped Pt(111) surfaces studied by HOPSTER et al. /91/. The oxygen adsorption probability increases exponentially with step density, whereas the probability for CO oxidation at step sites is lower. The overall oxidation yield depends thus on the relative occupation of step and terrace sites by oxygen atoms which, in turn, depends on the partial pressure ratio of oxygen and carbon monoxide in the catalytic reactor /91/.

Catalytic reactions on stepped surfaces involving hydrocarbons have been extensively studied by SOMORJAI and his co-workers /33,106,107,111-117/. According to BOUDART /3/ there are classes of structure-sensitive and structure-insensitive catalytic reactions. Reactions of the former type are expected to be influenced by the presence of steps and to depend on step density and most likely on step structure. We shall see that especially the breaking of molecular bonds such as C-H and C-C bonds are dependent on step structure. BLAKELY and SOMORJAI /112,113/ investigated the dehydrogenation reaction from cyclohexane to benzene and the hydrogenolysis of cyclohexane to n-hexane as a function of step and kink density on stepped platinum surfaces. The cyclohexane molecule (C_6H_{12}) consists of a ring of six carbon atoms like the benzene ring but with two hydrogen atoms bound to each carbon atom. The dehydrogenation reaction necessitates the breaking of six C-H bonds. The n-hexane molecule (C_4H_{12}) forms an aliphatic chain structure where two hydrogen atoms are bound to each carbon atom within the chain and three hydrogen atoms at the chain ends. The formation of n-hexane requires the breaking of C-C bonds.

The experiment was carried out at a constant hydrocarbon pressure of $4x10^{-8}$ Torr and a hydrogen to hydrocarbon pressure ratio of 20 for the dehydrogenation reaction and 300 for the hydrogenolysis reaction. The production rate of benzene and cyclohexane is given in terms of a turnover number, i.e., the number of product molecules per platinum surface atoms per second. Figure 4.8 shows the results reproduced from BLAKELY and SOMORJAI /113/. Part A depicts the dependence of the turnover number on step density. The hydrogenation reaction rate does not depend on step density (within the step density regime studied). However, the flat surface (step density equal to zero) yields a much smaller turnover number which indicates that the latter quantity must be dependent on step density but levels off already for rather small step density values (smaller than $1x10^{14}$ step atoms per cm^2). The hydrogenolysis turnover number increases linearly with step density. These findings show that C-H bond breaking (dehydrogenation) occurs readily on stepped surfaces and that the rate of C-C bond breaking is markedly increased with increasing step density. Part B of Fig. 4.8 illustrates that especially the presence of kinks within the steps causes the enhancement of the C-C bond breaking rate. The results shown in part B were obtained for stepped Pt(111) surfaces with the same step density of $2.5x10^{14}$ step atoms per cm^2 but with varying amounts of kink atoms per step (step directions rotated against the close-packed [110] direction by 7 and 20°, respectively). The hydrogenolysis turnover number increases linearly with kink density, which demonstrates that kinks appear to be very effec-

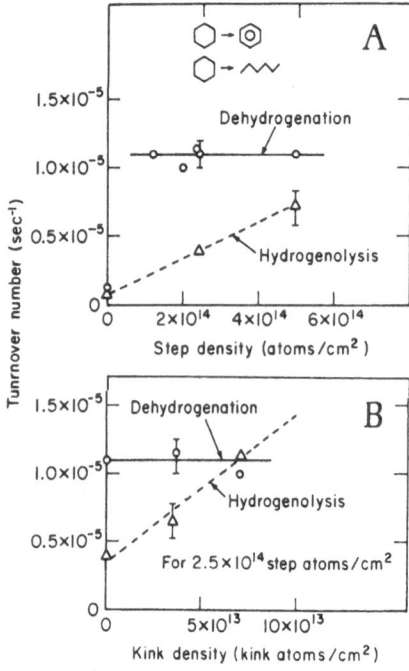

Fig. 4.8. Cyclohexane dehydrogenation to benzene (o) and hydrogenolysis to n-hexane (△) as a function of (A) step density and (B) kink density. The rates of hydrogenolysis per surface site are the slopes of the lines representing hydrogenolysis and are 2.5×10^{-4} molecules of n-hexane/kink atom sec and 2×10^{-5} molecules/ step atom sec. The slope is multiplied by 1.5×10^{-15} Pt atoms/cm^2 to obtain the desired units (according to BLAKELY and SOMORJAI /113/)

tive in breaking C-C bonds. Besides structural effects such as steps and kinks, reaction rates are also influenced by the presence of carbonaceous overlayers which develop during the dehydrogenation reactions /113/. The ordering of carbonaceous overlayers has also been found to depend on the presence of steps and especially kinks /107/.

The enhancement of catalytic reaction rates due to steps might be different for different substrate metals with otherwise equal structural properties. NIEUWENHUYS and SOMORJAI /115/ investigated hydrogenation and dehydrogenation reactions on Ir(111) and stepped Ir surfaces. The presence of steps does, for example, not increase the dehydrogenation rate on Ir by as much as observed on Pt, the reason being that C-H bond breaking occurs more easily on the flat Ir(111) surface than on the structurally equivalent Pt(111) surface. The catalytic activity of Au surfaces with respect to the reactions described above is completely negligible on either flat and stepped surfaces. The hydrocarbons do not even become chemisorbed under low-pressure conditions /33/. The comparison between Pt, Ir, and Au surfaces shows that marked effects on chemisorption and reaction rates caused by steps can only be expected if the respective processes occur at least to some extent already on the flat low index planes.

4.5 Atom and Molecule Scattering from Stepped Surfaces

Elastic scattering of atoms from low index metal surfaces occurs almost exclusively into the specular direction. There has been only one instance where first-order diffraction beams were observed with intensities comparable to the specular beam, i.e., the case of He scattering from the tungsten (112) plane /118/. This W plane displays a corrugated structure with close-packed atom rows in the [111] direction which are separated by a distance of $\sqrt{2}$ times the lattice constant. First-order diffraction beams appear when the incident He beam strikes the surface perpendicular to the close-packed atom rows and are absent if the direction of the rows lies in the plane of incidence. This example shows that the periodic interaction potential "seen" by the He atoms is rather smooth along the close-packed rows and significantly modulated in the direction perpendicular to them. Only recently BOATO et al. /119/ detected first- and second-order diffraction beams for He scattering on Ag(111) at low surface temperatures. In this case the diffraction intensities were of the order of 10^{-3} as compared to the specular beam. The Ag(111) exhibits a close-packed atom structure. The general absence of diffraction beams from low index metal surfaces may thus be attributed to the relatively small variation of the periodic interaction potential experienced by the scattering atoms.

Periodic step structures may be expected to cause more pronounced potential variations in the directions perpendicular to the steps as schematically indicated in Fig. 4.9. Recent investigations on thermal He beam scattering from stepped Pt surfaces by CEYER et al. /120/ and from stepped Cu surfaces by LAPU-JOULADE /121/ show that, indeed, the angular distribution of scattered atoms deviate markedly from just specular reflection.

The angular resolution in the experimental setup of CEYER et al. /120/ did not allow for separation of diffraction peaks but rather led to the characteristic features of the so-called rainbow scattering /122/. Figure 4.9 depicts two trajectories incident on the middle of the terrace and the step region, respectively, and pointing away from the macroscopic surface plane into two different directions. The sum of all possible trajectories leads to an angular distribution of the backscattered He atoms (rainbow distribution) which reflects the shape of the potential curve. The interference of scattered waves (wave mechanical description) from different terraces should result in the appearance of well-resolved diffraction beams. The angular rainbow distribution (scattering from a single terrace-step segment) then forms the envelope to the intensity distribution of the various diffraction beams.

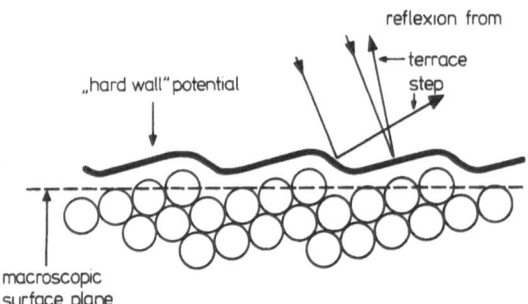

Fig. 4.9. Schematic representation of atom scattering from stepped surfaces. The trajectories drawn indicate backscattering into directions different from the specular direction with regard to the macroscopic surface plane (rainbow scattering)

CEYER et al. /120/ employed a thermal He beam of de Broglie wavelength $\lambda \sim 0.62$ Å and the stepped Pt surfaces Pt(S)-[5(111)x(111)] and Pt(S)-[9(111)x(111)]. The average terrace widths are 11 and 20 Å, respectively. The angular separation of neighboring diffraction peaks would therefore amount to 3 to 7° depending on diffraction order and angle of incidence. The divergence of the incident beam was 2° and the acceptance angle of the detector 7° /123/, making the separation of neighboring diffraction peaks impossible. Figure 4.10a reproduces the angular distribution of backscattered He atoms from the Pt(S)-[5(111)x(111)] surface for an angle of incidence of 45° and the incident He beam perpendicular to the steps. The specular reflection angle is marked by an arrow. Corresponding distribution curves obtained for the step direction within the plane of incidence (Fig. 4.10b) show only a broad scattering peak centered around the specular direction. In this case the total amount of He atoms backscattered within the plane of incidence is much smaller due to a large fraction of atoms scattered in out-of-plane directions.

LAPUJOULADE /121/ investigated He scattering from a Cu(S)-[3(100)x(111)] surface with a much higher angular resolution of 0.5^{o}. In this case the various diffraction peaks are well separated as shown in Fig. 4.11. It is interesting to note that the (20) diffraction order attains a very high intensity. This feature corresponds to

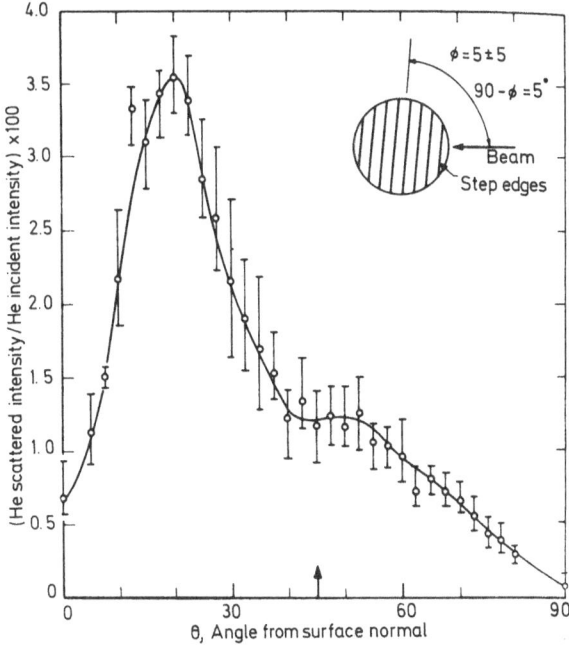

Fig. 4.10a. Angular distribution of backscattered He atoms from a Pt(S)-[5(111)x(111)] surface. Angle of incidence 45⁰ (according to CEYER et al. /120/). Incident He atoms perpendicular to steps

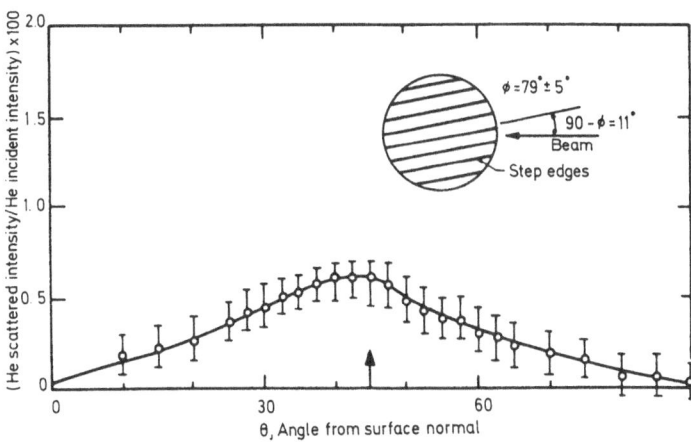

Fig. 4.10b. Angular distribution of backscattered He atoms from a Pt(S)-[5(111)x(111)] surface. Angle of incidence 45⁰ (according to CEYER et al. /120/). Step direction in plane of incidence

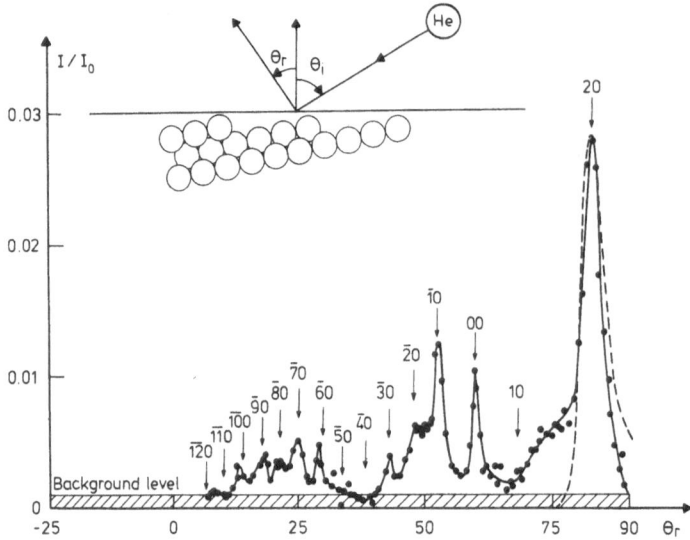

Fig. 4.11. He scattering from a Cu(S)-[3(100)x(111)] surface showing well-resolved diffraction peaks. Angle of incidence 60°. He beam perpendicular to steps as indicated in the inset (according to LAPUJOULADE /121/)

specular reflection from the (100) terraces. With the incident beam impinging at an angle of 60° to the macroscopic surface the specularly reflected beam from the (100) terraces should occur at an angle of 83° with regard to the macroscopic surface, as is indeed observed in Fig. 4.11. In this experiment the He beam impinges towards the open faces of the steps in contrast to the experiment with the stepped Pt surfaces (Fig. 4.10) which therefore exhibits the maximum intensity at a smaller reflection angle as schematically shown in Fig. 4.9.

The angular intensity distributions observed in Figs. 4.10 and 4.11 should yield valuable information on the interaction potential in the vicinity of the step edges. Recent theoretical treatments of atom scattering from periodic arrays were given by GARIBALDI et al. /124/, GARCIA /125/, GARCIA et al. /126/ and MASEL et al. /127,128/. With regard to the application to stepped surfaces and especially the experimental results described above, GARCIA and CABRERA /129/ reported a theoretical formalism which allows the exact solution of atom scattering from a hard corrugated surface. The characteristic features of the experimentally observed angular distribution (rainbow distribution, intensity of diffraction peaks) are well accounted for by this theory.

The scattering of diatomic molecules (H_2, D_2) from stepped Pt surfaces was investigated by BERNASEK and SOMORJAI /84,114/ and BERNASEK /123/. The angular distribution of the backscattered molecules was found to be much broader on the stepped surfaces as compared to the smooth low index Pt(111) surface. The broader distribution was interpreted in terms of a much more efficient thermal accommodation process of the incident molecules on the stepped surfaces. It was argued /123/ that transitions between rotational modes of the diatomic molecules during the scattering process are responsible for the energy transfer and thus for enhanced thermal accommodation and that they are more likely to occur on stepped surfaces. The differences in angular scattering distributions observed for H_2 and D_2 lend support to this explanation by considering the energy levels of the rotational modes of the two respective molecules and the phonon energies on the Pt substrate. An even better vehicle for accommodation (at least for momenta) provides an adsorbate layer of, for example, CO, on the Pt surface. In this case the internal degrees of freedom of the adsorbed molecule (bending modes) open additional paths for the energy transfer. Diatomic molecules but also atoms such as Ar are scattered from a CO covered Pt(111) surface with a cosine angular distribution characteristic of complete thermal accommodation /123/.

4.6 Surface Diffusion

Finally we touch on a further kinetic process involving adsorbed atoms or molecules on solid surfaces, i.e., surface heterodiffusion. We shall only consider in this context the altered behavior caused by regular step arrays.

Surface diffusion studies are often complicated by the simultaneous occurence of evaporation or bulk diffusion. BUTZ and WAGNER /130,131/ investigated the diffusion of palladium on tungsten. For this system the competitive processes are completely negligible in the investigated temperature range of 600-900 $^\circ$C and the Auger spectroscopy provides a sensitive technique for monitoring the Pd distribution over the studied W surfaces. Diffusion profiles, i.e., concentration versus distance plots, were obtained by using a scanning Auger microscope (SAM). The spreading of Pd proceeds via the formation of layers of distinct thickness /130/. The layer boundaries are found to advance with the square root of diffusion time. We shall not discuss the shape of the diffusion profiles (diffusion mechanism) any further but rather describe in more detail the anisotropic spreading behavior due to the presence of ordered step structures.

In order to investigate the diffusion anisotropy, Pd spots of about 100 μm in diameter and 15-20 monolayer thick were evaporated and functioned as diffusion sources. It is known /135/ that Pd forms above monolayer coverage three-dimensional clusters on W(110) at the temperature applied for the diffusion experiments. These

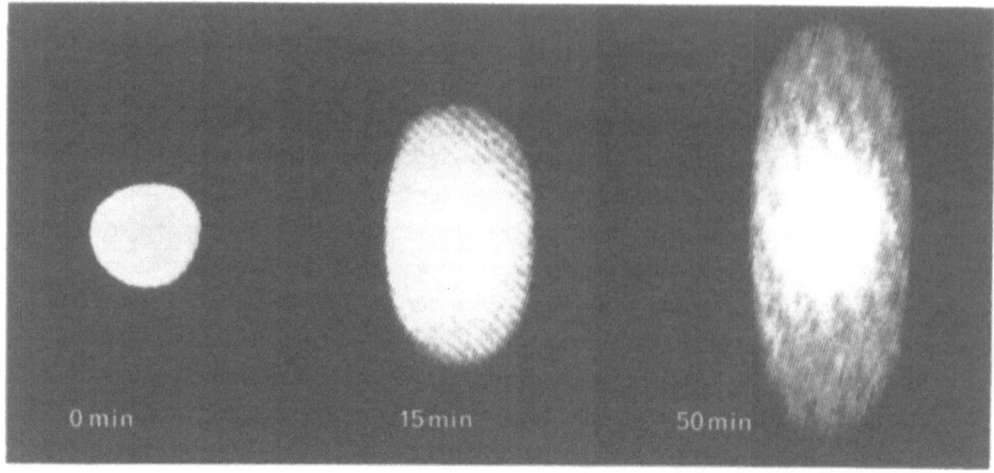

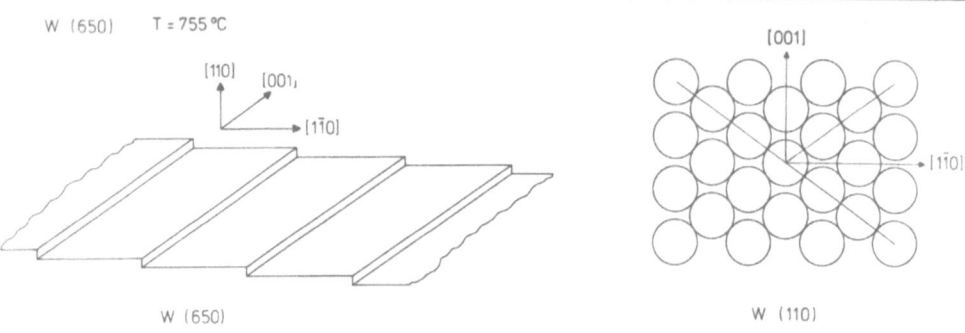

Fig. 4.12. Pd distribution on stepped W(650) after two heating times at 755 °C. Lower part: schematic of step structure and indications of crystallographic directions (BUTZ and WAGNER /130/)

clusters are therefore the actual diffusion sources. First studies on W(110) vicinals, i.e., W[20(110)x(1$\bar{1}$0)]and W[12(110)x(1$\bar{1}$0)], showed a pronounced spreading anisotropy represented by elliptical Pd distributions after the diffusion treatments. The long axis of the distribution, parallel to the step direction, was about 4 to 8 times larger than the short axis in the investigated temperature range. Despite the fact that the step density of the W[12(110)x(1$\bar{1}$0)] surface (7.4x10^6 cm^{-1}) is about twice as high as for the W[20(110)x(1$\bar{1}$0)] surface, the anisotropy turned out to be virtually the same. Figure 4.12 reproduces Pd distributions after two subsequent times at 755 °C for the W[12(110)x(1$\bar{1}$0)] surface.

Further investigations on curved W surfaces exhibiting (110) vicinals of continuously varying orientations were undertaken /131/ to study anisotropy effects for smaller step densities. This was accomplished by preparing a W surface of nearly ellipsoidal shape characterized by the two radii of curvature of 11 and

5 cm, respectively. The sample (6 cm in diameter) exposed the (110) plane near the central region and stepped portions with increasing density towards the sample edge. As derived from the measured radii of curvature, the maximum step densities near the sample edge amounted to $3x10^6$ cm^{-1} and $1.5x10^6$ cm^{-1} for the two respective sample directions. The corresponding terrace widths are 35 and 65 Å, respectively. LEED patterns obtained from the edge regions showed the characteristic spot splittings which complied with the terrace widths and step directions of the ordered step array expected for the macroscopic surface orientations. Onto this sample a regular array of about 180 circularly shaped Pd spots was evaporated. The sample was then heated for 5 min at 1000 K and the Pd distribution observed by SAM (Fig. 4.13).

The depicted image is composed of about 90 single photographs taken from the CMA oscilloscope display. The magnification chosen for a single picture provided an equal signal sensitivity over the scanned area. The initial Pd spots are still visible as bright circles. The elliptical areas around the bright spots represent the Pd diffusion zones. The long axes of the diffusion zones follow the step directions. The diffusion anisotropy characterized by the shape of the elliptical diffusion zones depends on both step density and step direction. The latter may be derived from the two crystallographic directions marked in the figure. Figure 4.13 reveals that the anisotropic spreading behavior is directly correlated with the step arrays and depends very sensitively on the presence of steps. Stepped regions with an average terrace width of larger than 400 Å exhibit already strong anisotropy effects. Such terrace widths correspond to inclination angles towards the (110) plane of less than 0.3°. Usual metallographic techniques allow the preparation of a particular surface orientation within $\pm$ 0.5°. It seems, therefore, almost impossible to obtain reliable informations on the diffusion anisotropy of singular low index planes without the disturbing action of steps.

In this context it was interesting to see whether the W(110) plane (twofold symmetry) exhibits by itself an anisotropy with regard to Pd diffusion. For this reason BUTZ and WAGNER /131/ employed a curved W surface with a larger radius of curvature of about 160 cm and yet succeeded in having the (110) plane on the surface. After heat treatment of the initial Pd spot pattern the diffusion zones in the (110) region were of circular shape within experimental errors. The directional anisotropy of the arrangement of W atoms in the close-packed (110) plane does therefore not give rise to corresponding anisotropy effects of the Pd diffusion. The observed radius of the diffusion zones within the (110) region was smaller than the long axis of the elliptical diffusion zones within the stepped surface regions.

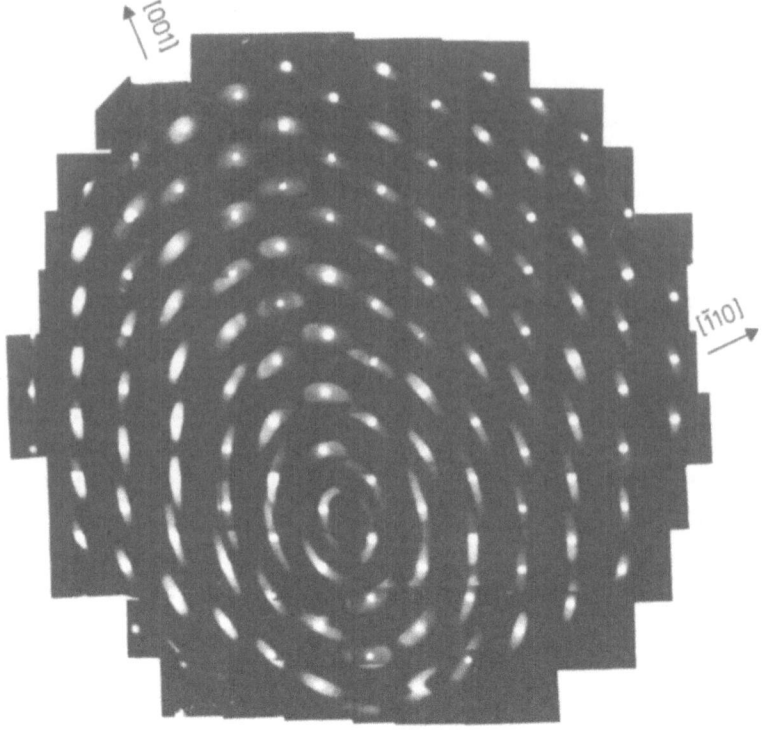

<u>Fig. 4.13.</u> Anisotropic spreading of Pd on a curved W surface observed with the scanning Auger microscope. The bright circular spots represent the initially de-posited Pd layers. The (110) region is located in the lower central part. The long axes of the Pd diffusion zones follow the step directions of the ordered step ar-ray (BUTZ and WAGNER /131/)

The latter finding together with the results of Figs. 4.12 and 4.13 shows that the spreading of Pd is markedly enhanced parallel to steps. The step edges seem to play a similar role in surface heterodiffusion as grain boundaries do in bulk dif-fusion. Further interpretations of the observed results may be taken form the ori-ginal paper /131/.

ROULET /132/ investigated the diffusion of Ag on the singular Cu(111) and Cu(100) faces as well as on two vicinal surfaces being rotated by 6° around the [$1\bar{1}0$] di-rection form the (111) and (100) face, respectively. The concentration of the dif-fusing Ag species was derived by measuring the oxidation rate of the Cu substrate which was found to depend on the Ag impurity concentration. The evaluation of the diffusion profiles yielded values of the preexponential and the activation energy of the Ag diffusion coefficients for various directions on the flat and the vicinal, i.e., stepped surfaces. As expected the diffusion coefficient turned out to be isotropic on the Cu(100) and Cu(111) planes but anisotropic on the stepped surfaces with the larger diffusion coefficient in the directional parallel to the steps. This anisotropy amounted, within the investigated temperature range 250 to 500 ^{o}C, to a factor 1.1 to 1.2 for the stepped Cu(100) and 2.4 to 1.7 for the stepped Cu(111) surface. The activation energy for diffusion was found to be lower for the stepped than for the flat surface. ROULET /132/ interpreted his results by assuming that diffusion occurs along ledges. In the case of the flat surfaces these ledges are distributed at random, whereas for the stepped surfaces additional ledges are present in one preferred orientation.

5. Conclusions

Vicinal surfaces provide an excellent means for producing regular step structures over macroscopic dimensions. Step density and step orientations can be adjusted at will by selecting the proper surface orientation. In most cases the regular step structures exhibit high thermal stability especially if the terraces are close--packed and the edges run in low index directions. The other alternative for the vicinal surface to minimize its free energy is rendered by a hill and valley structure formed by facets of neighboring low index planes.

The opportunity of having well-defined step structures available offers the possibility of studying systematically the influence of steps and their structural peculiarities on the physical properties of clean crystalline surfaces (e.g., electronic structure) and on their interaction with foreign molecules (e.g., adsorption kinetics, adsorption states, surface heterodiffusion, chemical reactions, etc.). Especially the latter prospect deserves considerable attention in view of the immense technological importance of these processes on real surfaces encountered in the field of corrosion, epitaxial growth and heterogeneous catalysis, to name only the most challenging areas of applied surface research. The understanding of the influences exerted by steps and the knowledge of the underlying physical principles can contribute valuable information for rationalizing observed processes on real surfaces such as polycrystalline films or the surfaces of small catalyst particles.

The study of step properties on vicinal surfaces of macroscopic dimensions is especially facilitated by the employment of a variety of surface specific tools currently used for studies on single crystal planes, such as LEED, AES, photoemission, and energy loss spectroscopy, work function measurements, etc..

In summarizing the results obtained so far, we can conclude that steps affect the electronic properties of crystal surfaces by decreasing the electron work function (at least for metals) and by changing the density of surface states or even causing new states in the case of semiconductors.

The influence of steps on the gas-solid interaction consists often in marked changes of the kinetic processes such as adsorption and reaction kinetics. Besides that, noticeable changes in binding states and adsorbate structures are observed. The most interesting effect of steps relates to the enhanced probability of breaking molecular bonds, e.g., dissociative adsorption of O_2, N_2, CO or breaking C-H and C-C bonds in hydrocarbons. Furthermore, the rate of dissociative desorption has been found to be influenced by steps. Atom and molecule scattering (elastic and inelastic) is noticeably affected by steps. Surface diffusion is a further area where strong effects due to steps have been found.

In some cases, changes in surface properties induced by steps have been quantitatively determined, in others only demonstrated more or less qualitatively. Further systematic studies, however, are necessary to elucidate in specific cases or for specific processes the physical reason for the observed effects in order to arrive at basic concepts which can be treated theoretically.

Acknowledgements

I like to thank E.A. Niekisch for stimulating this review article and the careful reading of the manuscript. Valuable discussions during the final preparation and fruitful comments of H.P. Bonzel, G. Comsa, H. Ibach and M. Henzler are gratefully acknowledged. I also like to thank my colleagues K. Besocke, R. Butz, B. Krahl--Urban and W. Erley who contributed important results and shared in enlightening discussions. Many of the drawings have been prepared and partly improved by Miss C. Damerow and Mrs. G. Flentje. Last not least I like to thank Miss H. Dohmen who did the typewriting and arranged with skill and patience the final version of the manuscript.

References

1 G.E. Rhead: Surf. Sci. 68, 20 (1977)
2 M. Henzler: Appl. Phys. 9, 11 (1976)
3 M. Boudart: Advanc. Catal. 20, 153 (1969)
4 W.K. Burton, N. Cabrera, F.C. Frank: Phil. Trans. Roy. Soc. London A243, 299 (1951)
5 C. Herring: Phys. Rev. 82, 87 (1951)
6 W. Kossel: Nachr. Akad. Wiss. Göttingen Math.-Phys. Kl. 1927, 135
7 I.N. Stranski: Z. Phys. Chemie 136, 259 (1928), 11, 421 (1931)
8 B. Lang , R.W. Joyner, G.A. Somorjai: Surf. Sci. 30, 440 (1972)
9 E.W. Müller, T.T. Tsong: Field Ion Microscopy (American Elsevier Publishing Comp., Inc., New York 1969)
10 J.W. May: Advances in Catalysis and Related Subjects 21, 152 (1970)
11 E. Bauer: in Topics in Applied Physics, 4: Interactions on Metal Surfaces (R. Gomer ed., Springer-Verlag Berlin, Heidelberg, New York 1975)
12 J.B. Pendry: Low Energy Electron Diffraction (Academic Press London, New York 1974)
13 H. Bethge, K.W. Keller, E. Ziegler: J. Crystal Growth 3/4, 184 (1968)
14 H. Bethge, K.W. Keller: J. Crystal Growth 23, 105 (1974)
15 K. Reichelt: J. Crystal Growth 35, 55 (1976)
16 G.A. Bassett: Phil. Mag. 3, 1024 (1958)
17 M. Henzler: Surf. Sci. 19, 159 (1970)
18 R. Butz, B. Krahl-Urban, E. Preuß, D. Bruchmann: phys. stat. sol. (a) 27, 205 (1975)
19 E. Preuß, B. Krahl-Urban, R. Butz: Laue-Atlas (Vieweg-Verlag Wiesbaden and Halsted Press New York, London, Toronto 1974)
20 E.W. Müller: Phys. Rev. 102, 618 (1956)
21 E.W. Müller, K. Bahadur: Phys. Rev. 102, 624 (1956)
22 W.P. Ellis, R.L. Schwoebel: Surf. Sci. 11, 82 (1968)
23 G.E. Rhead, J. Perdereau: in Colloque International sur la Structure et les Propriêtes des Surfaces des Solides (Editions due CNRS, Paris 1969)
24 L.G. Feinstein, M.S. Macrakis: Surf. Sci. 18, 277 (1969)
25 M. Henzler: Surf. Sci. 22, 12 (1970)
26 B.Z. Olshanetsky, S.M. Repinsky, A.A. Shklyaev: Surf. Sci. 69, 205 (1977)
27 M. Henzler: Surf. Sci. 36, 109 (1973)
28 J.W. Tester, C.C. Herrick, W.P. Ellis: Surf. Sci. 41, 619 (1974)
29 K. Besocke, H. Wagner: Surf. Sci. 52, 653 (1975)
30 W. Erley, H. Wagner: Surf. Sci. 74, 333 (1978)
31 L.C. Isett, J.M. Blakely: J. Vac. Sci. Technol. 12, 237 (1975)
32 H. Conrad, G. Ertl, E.E. Latta: Surf. Sci. 41, 435 (1974)
33 M.A. Chesters, G.A. Somorjai: Surf. Sci. 52, 21 (1975)
34 K. Besocke, B. Krahl-Urban, H. Wagner: Surf. Sci. 68, 39 (1977)
35 D.I. Hagen, B.E. Nieuwenhuys, G. Rovida, G.A. Somorjai: Surf. Sci. 57, 632 (1976)
36 R.L. Park, J.E. Houston: Surf. Sci. 18, 213 (1969)
37 J.E. Houston, R.L. Park: Surf. Sci. 21, 209 (1970)
38 M.J. Yacamán, T. Ocana Z.: J. Appl. Phys. 48, 418 (1977)
39 M. Henzler, J. Clabes: Proc. 2nd Intern. Cont. on Solid Surfaces, Japan, J. Appl. Phys. Suppl. 2, Pt. 2, 389 (1974)
40 Y.W. Tsang, L.M. Falicov: J. Phys. C: Solid State Physics 9, 51 (1976)
41 D.W. Blakely, G.A. Somorjai: Surf. Sci. 65, 419 (1977)
42 W.P. Ellis: Surf. Sci. 45, 569 (1974)
43 S.A. Isa, R.W. Joyner, M.W. Roberts: J. Chem. Soc., Faraday Trans. I. 72, 540 (1976)
44 G. Wulff: Z. Krist. 34, 449 (1901)
45 W.L. Winterbottom: in Surfaces and Interfaces, I-Chemical and Physical Characteristics, p. 133 (Syracuse University Press, Syracuse, New York 1967)
46 J.M. Blakely, R.L. Schwoebel: Surf. Sci. 26, 321 (1971)

47 P. Wynblatt: in Interatomic Potentials and Simulation of Lattice Defects, (Gehlen, Beeler and Jaffee ed., Plenum Press 1972, p. 633)
48 E.E. Gruber, W.W. Mullins: J. Phys. Chem. Solids 28, 875 (1967)
49 R.L. Schwoebel, E.J. Shipsey: J. Appl. Phys. 37, 3682 (1966)
50 R.L. Schwoebel: J. Appl. Phys. 40, 614 (1969)
51 H.J. Leamy, G.H. Gilmer: J. Crystal Growth 24/25, 499 (1974)
52 W.K. Burton, N. Cabrera: Discussions Faraday Soc. 5, 33 (1949)
53 N.A. Gjostein: Acta Met. 11, 957 (1963) I
54 N.A. Gjostein: Acta Met. II, 969 (1963) II
55 B. Krahl-Urban, E.A. Niekisch, H. Wagner: Surf. Sci. 64, 52 (1977) and references therein
56 J.C. Rivière: in Solid State Surface Science, 1: Work Function: Measurements and Results (Dekker New York 1969)
57 R. Smoluchowski, Phys. Rev. 60, 661 (1941)
58 N.D. Lang, W. Kohn: Phys. Rev. B 3, 1215 (1971)
59 K. Besocke, H. Wagner: Phys. Rev. B 8, 4597 (1973)
60 G.E. Ehrlich, F.G. Hudda: J. Chem. Phys. 44, 1039 (1966)
61 D.W. Basset, M.J. Parsley: J. Phys. D 3, 707 (1970)
62 W. Körner: phys. stat. sol. (a) 22, 523 (1974)
63 L.L. Kesmodel, L.M. Falicov: Solid State Comm. 16, 1201 (1975)
64 A. Huijser, J. van Laar: Surf. Sci. 52, 202 (1975)
65 M. Henzler: Surf. Sci. 25, 650 (1971)
66 W. Kuhlmann, M. Henzler: to be published
67 J.E. Rowe, S.B. Christman, H. Ibach: Phys. Rev. Lett. 34, 874 (1975)
68 P.E. Gregori, W.E. Spicer, S. Circaci, W.H. Harrison: Appl. Phys. Lett. 25, 511 (1974)
69 D.E. Eastman, J.L. Freeouf: Phys. Rev. Lett. 33, 1601 (1974)
70 V.T. Rajan, L.M. Falicov: J. Phys. C.: Solid State Phys. 9, 2533 (1976)
71 M. Schlüter, K.M. Ho, M.L. Cohen: Phys. Rev. B 14, 550 (1976)
72 E.W. Plummer, J.W. Gadzuk: Phys. Rev. Lett. 25, 1493 (1970)
73 R.B. Feuerbacher, B. Fitton: Phys. Rev. Lett. 29, 786 (1972), 30, 923 (1973)
74 P. Heimann, H. Neddermeyer, H.F. Roloff: Proc. of an Intern. Symp. Noordwijk, Netherlands 1976
75 H.P. Bonzel, C.R. Helms, S. Kelemen: Phys. Rev. Lett. 35, 1237 (1975)
76 M.C. Desjonquères, F. Cyrot-Lackmann: Surf. Sci. 53, 429 (1975)
77 M.C. Desjonquères, F. Cyrot-Lackmann: Solid State Comm. 18, 1127 (1976)
78 P.L. Jennings, G.S. Painter, R.O. Jones: Surf. Sci. 60, 255 (1976)
79 J.B. Salzberg, C.E.T. Concalves da Silva: Solid State Comm. 22, 207 (1977)
80 J.P. Hirth, G.M. Pound: Condensation and Evaporation (Pergamon Press Oxford 1963)
81 H.J. Leamy, G.H. Gilmer, K.A. Jackson: Surface Physics of Materials, Vol. 1 (J.M. Blakely ed., Academic Press New York 1975)
82 J. Perdereau, G.E. Rhead: Surf. Sci. 24, 555 (1971)
83 S.L. Bernasek, W.J. Siekhaus, G.A. Somorjai: Phys. Rev. Lett. 30, 1202 (1973)
84 S.L. Bernasek, G.A. Somorjai: J. Chem. Phys. 62, 3149 (1975)
85 H. Ibach, K. Horn, R. Dorn, H. Lüth: Surf. Sci. 38, 433 (1973)
86 K.E. Lu, R.R. Rye: Surf. Sci. 45, 677 (1974)
87 I.E. Wachs, R.J. Madix: Surf. Sci. 58, 590 (1976)
88 K. Christmann, G. Ertl: Surf. Sci. 60, 365 (1976)
89 N. Kasupke, M. Henzler: Verh. der DPG Münster 1977 and to be published
90 K. Besocke, S. Berger: Proc. 7th Intern. Vac. Congr. & 3rd Intern. Conf. on Solid Surfaces (Vienna 1977), Vol. II, p. 893
91 H. Hopster, H. Ibach, G. Comsa: J. Catal. 46, 37 (1977)
92 T. Engel, T. von dem Hagen, E. Bauer: Surf. Sci. 62, 361 (1977)
93 D.A. King: Proc. 7th Intern. Vac. Congr. & 3rd Intern. Conf. on Solid Surfaces (Vienna 1977), Vol. 1, p. 769 and references therein
94 D.L. Adam, L.H. Germer: Surf. Sci. 27, 310 (1971)
95 D.A. King, M.G. Wells: Proc. Roy. Soc. A 339, 245 (1974)
96 G. Pirug, G. Brodén, H.P. Bonzel: Proc. 7th Intern. Vac. Congr. & 3rd Intern. Conf. on Solid Surfaces (Vienna 1977), Vol. 2, p. 907
97 W. Erley: unpublished research (1977)

 98 H. Ibach: Surf. Sci. 53, 444 (1975)
 99 K. Zdansky, Z. Sroubek: J. Phys. F: Metal Phys. 6, L205 (1976)
100 D.M. Newns: Phys. Rev. 178, 1123 (1969)
101 W.H. Weinberg, R.P. Merrill: Surf. Sci. 33, 493 (1972)
102 L.C. Isett: Ph.D. Thesis, Cornell University (1974)
103 M.A. Chesters, M. Hussain, J. Pritchard: Surf. Sci. 35, 161 (1973)
104 R.H. Roberts, J. Pritchard: Surf. Sci. 54, 687 (1976)
105 K. Besocke et al., to be published
106 B.E. Nieuwenhuys, D.I. Hagen, G. Rovida, G.A. Somorjai: Surf. Sci. in print
107 K. Baron, D.W. Blakely, G.A. Somorjai: Surf. Sci. 41, 45 (1974)
108 R.J. Gale, M. Salmeron, G.A. Somorjai: Phys. Rev. Lett. 38, 1027 (1977)
109 G. Brodén, T.N. Rhodin, C.F. Brucker, R. Benbow, Z. Hurych: Surf. Sci. 59, 593 (1976)
110 M. Housley, R. Ducros, G. Piquard, A. Cassuto: Surf. Sci. 68, 277 (1977)
111 R.W. Joyner, B. Lang, G.A. Somorjai: J. Catal. 27, 405 (1972)
112 G.A. Somorjai, D.W. Blakely: Nature 258, 580 (1975)
113 D.W. Blakely, G.A. Somorjai: J. Catal. 42, 181 (1976)
114 S.L. Bernasek, G.A. Somorjai: Surf. Sci. 48, 204 (1975)
115 B.E. Nieuwenhuys, G.A. Somorjai: J. Catal. in print
116 G.A. Somorjai, R.W. Joyner, B. Lang: Proc. R. Soc. London A 331, 335 (1972)
117 D.R. Kahn, E.E. Petersen, G.A. Somorjai: J. Catal. 34, 294 (1974)
118 D.V. Tendulkar, R.E. Stickney: Surf. Sci. 27, 516 (1971)
119 G. Boato, P. Cantini, R. Tatarek: Proc. 7th Intern. Vac. Congr. & 3rd Intern. Conf. on Solid Surfaces (Vienna 1977), Vol. 2, p. 1377
120 S.T. Ceyer, R.J. Gale, S.L. Bernasek, G.A. Somorjai: J. Chem. Phys. 64, 1934 (1976)
121 J. Lapujoulade, Y. Lejay: Surf. Sci. 69, 354 (1977)
122 J.D. McClure: J. Chem. Phys. 52, 2712 (1970)
123 S.L. Bernasek: Ph.D. Thesis, University of California, Berkeley (1975)
124 U. Garibaldi, A.C. Levi, R. Spadacini, G.E. Tommei: Surf. Sci. 48, 649 (1975)
125 N. Garcia: Proc. Sol. Vac. Intern. Conf., Eindhoven, June 1976
126 N. Garcia, J. Ibánez, J. Solana, N. Cabrera: Solid State Comm. 20, 1159 (1976)
127 R. Masel, R. Merrill, W. Miller: Phys. Rev. B 12, 5545 (1975)
128 R. Masel, R. Merrill, W. Miller: J. Chem. Phys. 64, 45 (1976)
129 N. Garcia, N. Cabrera: Proc. 7th Intern. Vac. Congr. & 3rd Intern. Conf. on Solid Surfaces (Vienna 1977), p. 379
130 R. Butz, H. Wagner: Proc. 7th Intern. Vac. Congr. & 3rd Intern. Conf. on Solid Surfaces (Vienna 1977), p. 1289
131 R. Butz, H. Wagner: to be published
132 C.A. Roulet: Surf. Sci. 36, 295 (1973)
133 H. Wagner, K. Besocke: Berichte der Kernforschungsanlage Jülich, Nr. 1257 (1975)
134 K. Besocke, H. Wagner: Surf. Sci. 53, 351 (1975)
135 D. Paraschkevov, W. Schlenk, R.P. Bajpai, E. Bauer: Proc. 7th Intern. Vac. Congr. & 3rd Intern. Conf. on Solid Surfaces (Vienna 1977), p. 1737
136 N. Garcia: J. Chem. Phys. 67, 4304 (1977)

Applied Physics

A monthly journal

Board of Editors
S.Amelinckx, Mol. **V.P.Chebotayev,** Novosibirsk
R.Gomer, Chicago, IL., **H.Ibach,** Jülich
V.S.Letokhov, Moskau, **H.K.V.Lotsch,** Heidelberg
H.J.Queisser, Stuttgart, **F.P.Schäfer,** Göttingen
A.Seeger, Stuttgart, **K.Shimoda,** Tokyo
T.Tamir, Brooklyn, NY, **W.T.Welford,** London
H.P.J.Wijn, Eindhoven

Coverage
application-oriented experimental and theoretical
physics:

Solid-State Physics   *Quantum Electronics*
Surface Sciences   *Laser Spectroscopy*
Solar Energy Physics   *Photophysical Chemistry*
Microwave Acoustics   *Optical Physics*
Electrophysics   *Integrated Optics*

Special Features
rapid publication (3-4 months)
no page charge for **concise** reports
prepublication of titles and abstracts
microfiche edition available as well

Languàges
mostly English

Articles
original reports, and short communications review
and/or tutorial papers

Manuscripts
to Springer-Verlag (Attn. H.Lotsch), P.O.Box 105 280
D-6900 Heidelberg 1, F.R. Germany

Springer-Verlag
Berlin
Heidelberg
New York

Place North-American orders with:
Springer-Verlag New York Inc., 175 Fifth Avenue,
New York, N.Y. 100 10, USA

M.A. van Hove, S.Y. Tong

Surface Crystallography by LEED

Theory Computation and Structural Results

1979. 19 figures, 2 tables. Approx. 300 pages
(Springer Series in Chemical Physics,
Volume 2)
ISBN 3-540-09194-7

Contents:
Introduction. – The Physics of LEED. Basic
Aspects of the Programs. – Symmetry and Its
Use. – Calculation of Diffraction Matrices for
Single Bravais-Lattice Layers. – The Com-
bined Space Method for Composite Layers:
Matrix Inversion. The Combined Space
Method for Composite Layers: Reverse Scat-
tering Perturbation. – Stacking Layers by
Layer Doubling. – Stacking Layers by Renor-
malized Forward Scattering (RFS) Pertur-
bation. – Assembling a Program: The Main
Program and the Input. – Subroutine Lis-
tings. – Structural Results of LEED Crystallography. – Appendices.

Monte Carlo Methods in Statistical Physics

Editor: K. Binder

1979. 91 figures, 10 tables. XV, 376 pages
(Topics in Current Physics, Volume 7)
ISBN 3-540-09018-5

Contents:
K. Binder: Introduction: Theory and "Techni-
cal" Aspects of Monte Carlo Simulations. –
D. Levesque, J.J. Weis, J.P. Hansen: Simulation
of Classical Fluids. – *D.P. Landau:* Phase Dia-
grams of Mixtures and Magnetic Systems. –
R. Müller-Krumbhaar: Simulation of Small
Systems. – *K. Binder, M.H. Kalos:* Monte Carlo
Studies of Relaxation Phenomena: Kinetics of
Phase Changes and Critical Slowing Down. –
H. Müller-Krumbhaar: Monte Carlo Simu-
lation of Crystal Growth. – *K. Binder,
D. Stauffer:* Monte Carlo Studies of Systems
with Disorders. – *D.P. Landau:* Applications in
Surface Physics.

Interactions on Metal Surfaces

Editor: R. Gomer

1975. 112 figures. XI, 310 pages
(Topics in Applied Physics, Volume 4)
ISBN 3-540-07094-X

Contents:
J.R. Smith: Theory of Electronic Properties of
Surfaces. – *S.K. Lyo, R. Gomer:* Theory of
Chemisorption. – *L.D. Schmidt:* Chemi-
sorption: Aspects of the Experimental Situ-
ation. – *D. Menzel:* Desorption Phenomena. –
E.W. Plummer: Photoemission and Field
Emission Spectrocopy. – *E. Bauer:* Low Ener-
gy Electron Diffraction (LEED) and Auger
Methods. – *M. Boudart:* Conepts in Hetero-
geneous Catalysis.

Electron Spectroscopy for Surface Analysis

Editor: H. Ibach

1977. 123 figures, 5 tables. XI, 255 pages
(Topics in Current Physics, Volume 4)
ISBN 3-540-08078-3

Contents:
H. Ibach: Introduction. – *D. Roy, J.D. Carette:*
Design of Electron Spectrometers for Surface
Analysis. – *J. Kirschner:* Electron-Excited Core
Level Spectroscopies. – *M. Henzler:* Electron
Diffraction and Surface Defect Structure. –
B. Feuerbacher, B. Fitton: Photoemission Spec-
troscopy. – *H. Froitzheim:* Electron Energy
Loss Spectroscopy.

Springer-Verlag
Berlin
Heidelberg
New York